THE ECONOMIC GEOGRAPHY OF THE TOURIST INDUSTRY

In a variety of settings, from remote villages in less developed countries to major metropolitan regions in advanced societies, policymakers are enthusiastically encouraging the development of tourist-oriented facilities in the hope of generating economic growth. Yet, this potentially vast economic resource has received little attention to date.

The Economic Geography of the Tourist Industry explores how tourism is defined and examines whether or not tourism can be conceptualized as an industry. Detailed analyses of key sectors of the travel and tourism industry, such as tour operators, airlines and the hotel industry, are backed by a broad range of international case studies. The book also investigates issues such as business cycles, labour dynamics, entrepreneurship, and the role of the state in tourism and concludes that the production of tourism-related services has characteristics commonly associated with 'harder' production sectors, such as manufacturing and producer services.

The Economic Geography of the Tourist Industry bridges the gap between tourism research and economic geography by bringing together leading academics in geography, planning and tourism. This unique interdisciplinary approach offers an agenda for research which will help enrich the economic geography of travel and tourism.

Dimitri Ioannides is an assistant professor of Community and Regional Planning at Southwest Missouri State University. **Keith G. Debbage** is an associate professor of Geography at the University of North Carolina at Greensboro.

THE ECONOMIC GEOGRAPHY OF THE TOURIST INDUSTRY

A supply-side analysis

160101

Edited by
*Dimitri Ioannides and
Keith G. Debbage*

London and New York

First published 1998
by Routledge
11 New Fetter Lane, London EC4P 4EE

Simultaneously published in the USA and Canada
by Routledge
29 West 35th Street, New York, NY 10001

Typeset in Garamond by
Keystroke, Jacaranda Lodge, Wolverhampton
Printed and bound in Great Britain by
Biddles Ltd, Guildford and King's Lynn

British Library Cataloguing in Publication Data
A catalogue record for this book is available from the British Library

Library of Congress Cataloguing in Publication Data
The economic geography of the tourist industry : a supply-side
analysis / [edited by] Dimitri Ioannides and Keith G. Debbage.
p.   cm.
Includes bibliographical references (p.   –) and index.
1. Tourist trade. I. Ioannides, Dimitri, 1961–   . II. Debbage, Keith G., 1959–
G155.A1E25   1998
338.4′791—dc21 97–40285

ISBN 0–415–16411–7 (hbk)
ISBN 0–415–16412–5 (pbk)

In memory of
Billy Lee Lambert
(1 December 1930–27 April 1996)
and
Alexander Ioannides
(3 September 1972–8 May 1993)

CONTENTS

CONTENTS

CONTENTS

ILLUSTRATIONS

Plates

Figures

TABLES

NOTES ON CONTRIBUTORS

Beauregard, Robert is a professor in the Milano Graduate School of Management and Urban Policy at the New School for Social Research in New York City, USA. He writes on urban issues, focusing on urbanization and economic development. His latest book is *Voices of Decline: The Postwar Fate of US Cities* (Blackwell, 1993).

Daniels, Peter is Professor of Geography and Director of the Service Sector Research Unit, University of Birmingham, UK. His research interests include services in the global economy with particular reference to service MNEs, producer services and urban and regional development, and the role of business services in local economic restructuring. He has published widely on the geography of services, including *Service Industries in the World Economy* (Blackwell, 1993) and edited, with J. R. Bryson, *Service Industries in the Global Economy, Vol. 1: Service Theories and Service Employment, Vol 2: Services, Globalization and Economic Development* (Elgar, 1998)

Debbage, Keith G. is an associate professor of Geography at the University of North Carolina at Greensboro, USA. His research interests include the international airline industry and hub-and-spoke networks, the tourist industry and resort cycles, and processes of urban-economic restructuring. He consults regularly with local cities and regional agencies, and has prepared several land-use corridor studies for local jurisdictions.

Gill, Kara completed a Masters of Arts Degree in Economic Geography at McGill University, Montreal, Canada in 1997. She currently works as Project Administrator in the Control Systems and Simulators Division of CAE Electronics. She has remained active in the field of tourism supplier promotion and is currently developing training/travel packages for overseas and domestic recreational pilots.

Hall, C. Michael is a researcher at the Centre for Tourism, University of Otago, Dunedin, New Zealand. Previous appointments have included the Victoria University of Canberra and the University of New England,

Australia, and Massey University, New Zealand. He is the author, co-author or co-editor of fourteen books and over two hundred other publications in the tourism, environmental history and heritage management fields. Current research includes various aspects of tourism planning and development, especially in peripheral areas, and the production of tourism knowledge.

Haywood, Michael is a professor in the School of Hotel and Food Administration, University of Guelph, Ontario, Canada. He teaches courses in tourism, strategic management, and service operations. In 1992 he received the John Wiley and Sons award for lifetime contributions to hospitality and tourism research. He serves on the editorial boards of six academic and industry trade journals and is an active researcher and consultant with interests in urban, rural, and island tourism, the strategic management of tourism enterprises, and management development programs.

Ioannides, Dimitri is an assistant professor of Community and Regional Planning in the Department of Geography, Geology, and Planning at Southwest Missouri State University, USA. He is also Coordinator of the Undergraduate Program in Planning and serves as Director of the Graduate Program in Resource Planning. His research focus is on the supply-side of tourism, tourism planning, and issues relating to sustainable development.

Mackun, Paul J. is a doctoral candidate (Ph.D. expected in spring 1998) in the Department of Geography at the University at Buffalo, USA. His research interests are in industrial geography, population geography, and tourism.

Milne, Simon is an associate professor in the Department of Geography at McGill University, Montreal, Canada. He also coordinates the McGill Tourism Research Group – a multidisciplinary grouping of faculty and graduate students. Simon's current research focuses on the links between tourism and the development process in urban and rural community settings. He is presently involved in tourism research in Russia, Cuba, Canada, New Zealand and in microstates of the South Pacific.

Pohlmann, Corinne completed a Master of Arts Degree in Economic Geography at McGill University, Canada in 1994. Since then, she has worked for ACTA (The Alliance of Canadian Travel Associations) and is currently the National Manager of Membership Services at their head office in Ottawa, Canada. She also works part time as a tourism geography instructor at Algonquin College in Ottawa.

Roehl, Wesley is an associate professor in the William F. Harrah College of Hotel Administration at the University of Nevada, Las Vegas, USA. His

FOREWORD

Twenty years ago, as a graduate student and aspiring economic geographer, I frequently lamented the dearth of quality analysis of tourism, an activity that had significant – and growing – impact on national and regional economies, landscapes, environments, and cultures. In discussions with geographer-colleagues, we speculated on the reasons. Perhaps it was because the service economy was still a relatively new concept, and economic geographers and other scholars were still oriented to primary industries and manufacturing, with only minor focus on retailing and other tertiary activity. Or maybe it was the paucity and reliability of statistical indicators, an obvious hindrance in the 1970s, when many geographers were seized by the 'quantitative revolution'. Some, myself included, reckoned that because tourism was a leisure activity, even fun, scholars regarded it as inherently illegitimate. What Puritanism!

In the intervening two decades, the tourism industry has boomed beyond the most optimistic projections of the 1960s and 1970s. For example, air transportation, a major component of long-distance tourism, has roughly doubled in twenty years, with substantial numbers of routes oriented solely or dominantly to leisure traffic (every Saturday last winter, American Airlines alone operated twelve Boeing 757s into Vail/Eagle, Colorado, delivering more than 2,000 skiers in a single day). In some resort and recreation destinations, double-digit annual growth has been the norm. Entire sections of US downtowns, shorn of their historic functions in the post-war era and abandoned in the 1960s and 1970s, have been recycled to serve the meeting and convention industry, as well as 'tourists' visiting from nearby suburbs. And the tourism industry world-wide has effectively become an enormous geography teacher, as millions of people learn – and often mislearn – about places and peoples through tour escorts at the front of a sightseeing bus, the plethora of guidebooks, and other instructors. In short, the impact of tourism on geography has become quite enormous.

Although a reasonable amount of quality scholarship on tourism has been produced in recent years, the volume of work has by no means kept pace with the growth of the industry. Simply put, there have not been and certainly are no longer any valid reasons to exclude the study of tourism from the broader literature on the geography of economic activity, and the legitimacy and vitality of economic geography as a sub-discipline will be undermined by geographers' inattention to tourism. This book will help address this obvious deficit.

Robert A. Britton
Managing Director – Corporate Communications
American Airlines

1

INTRODUCTION: EXPLORING THE ECONOMIC GEOGRAPHY AND TOURISM NEXUS

Dimitri Ioannides and Keith G. Debbage

> The geographic study of tourism requires a more rigorous core of theory in order to conceptualize fully its role in capitalist accumulation, its economic dynamics, and its role in creating the materiality and social meaning of places. . . . Geographers conducting research on tourism have the potential to integrate their topic into the forefront of contemporary debates in the discipline.
>
> (S. Britton 1991: 452)

Background

In the late twentieth century the pursuit of leisure has become an essential ingredient of modern consumer culture (Featherstone 1991). Nowadays, particularly in western societies, people regard leisure and recreation as their right, an avenue of escapism from the monotonous routine of everyday life on the job. Travel and tourism are key representations of the growing dichotomy between work and leisure. Becoming a tourist allows a person to move temporarily away from the usual place of residence and visit destinations offering views and experiences which do not feature within their ordinary everyday lives. These views and experiences, which Urry (1988) has dubbed 'the tourist gaze', are sought predominantly for the purpose of deriving pleasure. Even business travelers, who in certain destinations may represent a significant proportion of all tourists, commonly participate in various leisure and recreation-oriented activities during their trips (Shaw and Williams 1994).

In the past, the experiential aspects associated with leisure-based activities such as travel and tourism have led a number of observers from a variety of disciplines to dismiss the study of leisure and tourism as frivolous, especially when compared to seemingly more 'important' academic research on such

1

matters as manufacturing and related issues (Blank 1989; R. Britton 1978; Hall 1994). Krippendorf appropriately mentions that most people do not regard leisure and tourism-related actions as 'serious' since these cannot be equated to work, 'and only work can be serious' (1987: ix).

Fortunately it appears that the tide is gradually changing, as in recent years, more and more social scientists have embraced the study of travel and tourism. To an extent, this rising interest in tourism reflects the recognition that leisure-related activities are indeed important functions within modern societies and that their consumption 'cannot be separated from the social relations in which they are embedded' (Urry 1995: 129). Researchers have begun to realize, for example, that through studying tourism they can gain a better grasp on issues relating to economic or societal restructuring. Tourism and leisure are related to broader economic and societal issues including the globalization of culture, the commodification of place, and territorial competition (Featherstone 1991; Harvey 1989a; Shaw and Williams 1994). Within this realm, tourism itself is now commonly regarded as part of the nascent 'culture industry' and, as such, is an experience which can be bought and sold as a commodity just like other household recreation-oriented products (e.g. televisions, stereos, books, magazines, etc.). Similar to other consumer goods, travel and tourism activities are capitalistically organized, involving the production of both intangible/experiential, but also tangible, goods and services.

Importantly, certain types of travel and tourism activities (e.g. hotels and catering establishments) cannot be considered purely consumer services since they also function as producer or support services (Daniels 1993). For example, hotels not only meet consumer-based or final demand through their experiential products, but they also satisfy intermediate demand (e.g. through information exchange and the development of business contacts) since they also supply other businesses and government institutions with convention facilities which in turn can deliver additional service-related outputs. A clear example of the producer service component of tourism is the rapidly growing business travel sector (Bull and Church 1994), much of which has little to do with either pleasure or leisure.

According to S. Britton (1991: 453–54), there now exists a complex ensemble of 'enterprises, industries, markets, state agencies' – a travel and tourism production system – whose overriding aim is to market not just the means to an end, but the end itself, the travel experience (see also Urry 1995). The sheer volume of this production system, which according to numerous sources now forms the world's largest industrial complex (Lundberg et al. 1995), is a particularly compelling rationale for the growing academic interest in travel and tourism. The travel and tourism sector now employs almost 7 per cent of the global workforce and accounts for just over 6 per cent of gross domestic product (GDP) world-wide (Apostolopoulos et al. 1996; C. M. Hall 1994; Theobald 1994). In a variety of settings, from remote villages in less developed countries to major metropolitan regions in advanced regions, policymakers enthusiastically pursue the development of tourist-oriented facilities (e.g. airports, convention centers, sports stadia, aquaria, museums) in the hope of attracting foreign exchange and, in turn, generating economic growth, creating jobs, and eventually diversifying the economy. Even though tourism has been subjected to increasing criticism in recent years, due to the environmental and sociocultural costs it imposes on host societies and the realization that it is not as 'clean' an industry as once thought, to a major extent the attitude of most policymakers towards this sector has remained overwhelmingly favorable (Judd 1995).

Given the sector's magnitude and high visibility in so many localities around the globe, it is hardly surprising that a growing number of social scientists have now jumped on the bandwagon of tourism-related research. In the past two decades alone, there has been an explosion of travel and tourism-related literature in the form of books, journal articles, and monographs. Anthropologists (V. Smith 1977), economists (Gray 1970; Lundberg *et al.* 1995), geographers (R. Britton 1978; S. Britton 1991; Butler 1980; Shaw and Williams 1994), sociologists (Cohen 1995; Lanfant 1993; Urry 1990), and even political scientists (C. M. Hall 1994; Judd 1995; Richter 1989) and urban planners (Gunn 1994; Inskeep 1988) have explored tourism within their respective disciplines, and their writings have certainly enriched our understanding of the sector.

Geographic research in travel and tourism

In recent years, geographers have been particularly active in tourism research. Borrowing viewpoints from other disciplines and integrating these with their own spatial perspective, geographers have actively explored and sought to interpret a diversity of research problems relating to travel and tourism. Many geographic texts on tourism, leisure, and recreation devote at least some attention to the definitional complexities related to the travel sector, and tourists themselves. Like most other social scientists, geographers are intrigued by tourism's economic, sociocultural, and environmental impacts in a variety of settings (R. Britton 1978; Mathieson and Wall 1982; Pearce 1989 and 1995; Zurick 1992). They have paid special attention to tourism's land-use implications in receiving areas, examining ways by which the sector can be developed in order to maximize economic returns while minimizing negative cultural or environmental externalities (Pearce 1995; Wilkinson 1989). Within this context, studies have explored the issues of tourism carrying capacities and sustainable tourism development (Butler 1980; Williams and Gill 1994). There has also been research on the patterns of tourist flows and the spatial proclivities of tourism-related activities (Pearce 1995; Williams and Zelinsky 1970). Furthermore, a substantial body of geographic literature relates to the dynamics of tourism evolution in destination areas (Butler 1980; Christaller 1963; Debbage 1990; Haywood 1986; Ioannides 1992; Miossec 1977).

Despite the surge of geographic interest in tourism, a number of critics have lamented the descriptive nature of most studies in the discipline. As S. Britton (1991) asserts, the mainstream geography journals (e.g. *Area, Economic Geography, Progress in Human Geography, Urban Geography*) have generally ignored much of this body of tourism research, precisely because of its 'narrow scope and shallow theoretical base' (475). Of course this dearth of rigorous theory characterizes, albeit to varying degrees, tourism research in much of the social sciences and has been a major hurdle in terms of advancing 'our understanding of the processes contributing to tourism (i.e. its growth, development, operation, and management) and the way it functions in different environments' (Page 1995: 3). Page explains that the lack of a solid conceptual framework largely relates to the way that tourism research has evolved as a sub-discipline of a variety of other disciplines (e.g. economics, sociology), meaning it is missing the theoretical cohesiveness normally associated with most major research fields in the social sciences.

A particular gap in the literature relates to the supply-side of tourism, a peculiar phenomenon when considering the industry's apparent magnitude and importance in numerous localities (S. Britton 1991; Ioannides 1995; Sinclair 1991). True, a number of studies examine the organizational and institutional aspects of the tourism sector (Burkart and Medlik 1981; Gee *et al.* 1989; Mill and Morrison 1985; Pearce 1989; Williams and Shaw 1988) yet, once more, these are largely descriptive and have been developed outside a rigorous theoretical framework. For example, the political economy perspective that has characterized the recent work of many regional economists and geographers is largely absent in examinations of the travel and tourism sector (though see S. Britton 1991; Shaw and Williams 1994).

The silence of economic geographers

It is no secret that most geographers who engage in locational studies of economic activity have historically focused disproportionately on manufacturing activities. Fortunately, it now seems that this myopic vision of the space economy has begun to abate somewhat given recent transformations in the production system and the structural changes in the production of goods and services. Thus, since the mid-1980s there has been a major revolution in economic geography, as an increasing number of academics have begun paying attention to the spatial patterns of the rapidly expanding service sector (Christopherson 1993; Coffey and Bailly 1992; Daniels 1993; Dunham 1997; Malmberg 1994; Townsend 1991; Urry 1987).

Nevertheless, despite the surge of interest in services, most studies concentrate heavily on the producer services whereas consumer services (perhaps with the exception of retailing) have been neglected. Tourism and leisure-related services have aroused only sporadic interest in the geography of production (Shaw and Williams 1994) reflecting their present status as the 'poor cousin' or 'Cinderella' of services (Ioannides 1995: 58). A cursory glance at the table of contents of most mainstream economic geography texts and periodicals reveals almost no reference to tourism and travel-related services (Berry *et al*. 1993; de Souza and Stutz 1994; Chapman and Walker 1991). The few books mentioning travel and/or tourism do so as an afterthought (Hartshorn and Alexander 1988; Healy and Ilbery 1990), and lack any theoretical discussion on the sector's spatial proclivities or the dynamics of its labor markets. Importantly, a number of service-oriented texts also overlook the role of consumer services like tourism. For instance, while Daniels (1993) briefly examines the British airline sector within the overall scope of tradable producer services, he does not indicate the broader interconnections of this sector with the much larger tourism production system. Moreover, though certain academics devote considerable attention to the role of services in the restructuring of cities, they surprisingly omit any discussion of tourism's significance as a job and income generator for metropolitan economies (Stanback and Noyelle 1982). Finally, even though in recent years, an increasing number of economic geographers have advocated a more comprehensive 'services-informed' research agenda for producer, consumer and public-sector oriented services (Dunham 1997; Townsend 1991), there have been no such holistic approaches relating to the study of tourism.

Reasons behind the silence

Economic geographers have traditionally shunned the study of tourism because, like many other academics, they regard it and, for that matter, most consumer services as subordinate to more 'acceptable' or 'serious' activities like manufacturing or producer services (R. Britton 1978; Pearce 1989). In the United States, such an attitude most likely reflects allegiance to the nineteenth-century Protestant work ethic which stresses 'tangible, physical construction and production' (Blank 1989: 2). Certain authors have even criticized the physical regeneration of localities for the consumption of tourism as an affront to the hard, concrete labor of yesteryear (Boyer 1992). Such attitudes, unfortunately, mirror the stigma tourism has to bear as an activity traditionally associated with pleasure. This is despite growing recognition that certain components of the travel and tourism sector (e.g. business travel) can only tangentially be associated with pleasure-related activities or traditional consumer-based services.

It is, however, not altogether surprising that such misconceptions hound tourism research within the sphere of economic geography. After all, academics generally have overriding reservations concerning the importance 'of services in local economic development' (Townsend 1991: 309). A prevailing perception is that sectors such as manufacturing or producer services are the foundation of wealth creation in any community whereas consumer services such as tourism, which often depend on unskilled, highly feminized, low-wage, seasonal and/or part-time labor, are merely peripheral actors. Moreover, the tourism sector's poor image derives from accusations that it makes limited use of technologies and repeatedly demonstrates low productivity levels. A common charge is that the sector offers few opportunities for tax-base expansion and 'because wages are minimal and turnover is high, it is likely that workers will have trouble supporting their families and will require welfare services from the government' (Judd 1988: 393).

As is often discussed throughout this text, a parallel reason for tourism's subservient role compared to other economic activities, like manufacturing and producer services, derives from its 'invisible condition'. As Ashworth argues, 'a tourism industry does not exist in the same institutional sense as does the motor car industry' (1992: 4). Rather 'the industry itself is diverse and rambling – sleeping bag manufacturers and art museums claim membership in it' (R. Britton 1978: 17). The tourism industry is multifaceted comprising the aggregate of goods and services produced by different economic sectors, such as lodging, transportation, retailing, or recreation facilities. Thus, it is exceedingly hard for researchers to estimate with any degree of accuracy the sector's magnitude or significance in terms, for example, of generated revenues or employment.

Tourism's status as an industry remains a hotly debated issue. Some analysts argue that like any other industry there is a need to view tourism as a totally industrialized sector, that is from a supply-side perspective. This approach means tourism has to be defined in terms of the goods and services it produces (Smith 1988). Still, others maintain that the industrial approach is shortsighted since only a portion of the sector is industrialized (G. Hughes 1991). When tourists visit a destination they do not depend entirely on so-called tourism industries (e.g. accommodation, tour guiding services) but also commonly participate in 'non-market' activities (e.g. independent sightseeing, photography, hiking) (Leiper 1990a: 603).

Moreover, as Leiper argues, a considerable proportion of a locality's workforce can exist regardless of tourism (e.g. a restaurant or a gift shop) and may directly or indirectly cater to tourist needs. To further complicate matters, a number of public/quasi public organizations (e.g. national tourism administrations, visitors' bureaux, non-profit groups) engage in the business of tourism. Thus, beyond the collection of private-sector industries and other commercial activities, the production and consumption of tourism experiences also depend on public agencies, host communities, and individuals. It is precisely this 'non-industrialized' portion of tourism that contributes to the sector's 'poor reputation in the eyes of policy analysts, government officials, economic analysts and industry leaders not involved in tourism' (S. Smith 1988: 182).

The economic geography of tourism: towards a specialist approach

Unfortunately, the weak level of theorization associated with most investigations relating to tourism's supply-side means that we have been unable to strengthen our understanding of the sector's workings and resulting socio-economic actions. For instance, there is a pressing need to obtain answers to such questions as: What are the dynamics of tourism's labor markets? What is the role of entrepreneurship in the tourism industry? How does the diffusion of information technologies influence the travel sector? What are the effects of spatial and temporal contingencies on travel-related enterprises? Why do certain localities benefit more than others from tourism? How do the sector's various actors and stakeholders interact to promote destinations and what is the role of state institutions in this process? How do alleged economic or societal changes, particularly the emergence of post-modernism and the advent of post-Fordist modes of production and consumption, play out in tourism? Is Scott (1993) correct when he points out that tourist areas feature characteristics normally identified in flexible production agglomerations?

To begin answering these questions, there is a strong need to broaden the scope of geographic research in tourism. This can be achieved not only by further legitimizing tourism and recreation research as a geographic sub-discipline in its own right, but also by embracing this study area within a number of other sub-disciplines within geography. For example, various branches of geography can benefit from incorporating travel and tourism studies within their realm. As R. Britton appropriately argued almost two decades ago

> Tourism is a phenomenon variably distributed in space (and time), and it can thus be approached from a variety of geographical branches. The locations of markets and destinations, and the flow of people, capital, goods, and ideas are at the core of tourism. It influences the form, use, and protection of the landscape. Cultural values such as changing tastes in accommodation transform rural and urban landscapes. . . . The obvious impact of tourism on the physical environment . . . call for attention by physical geographers. . . .
>
> (R. Britton 1979: 278)

Tourism also belongs to the domain of a variety of other geographic sub-disciplines such as cultural, transportation, urban, and political geography. Referring to the study of urban tourism, for example, Ashworth (1992) and Page (1995) have indicated the need for both theoretical and empirical approaches which can strengthen our understanding of tourism functions within cities.

The argument can, therefore, also be made for allowing the study of tourism to 'float' within the sphere of economic/industrial geography. Its position within this particular sub-discipline derives in part from its status 'as a commercial activity' (R. Britton 1979: 279). Just like more 'conventional' sectors such as manufacturing, producer services, or retailing, travel-related activities are subject to locational dynamics, product cycles, processes of industrial concentration, economies of scale and scope, labor and entrepreneurial issues, the globalization of capital, and economic restructuring. Such issues are well-established topics of interest within economic geography (Berry *et al.* 1993). Thus, why should economic geographers not extend these research issues to the enormous tourism production system, especially given its large and rapidly growing employment base?

The conceptual and methodological complications associated with tourism can no longer be thought of as legitimate reasons for avoiding the study of this sector from an economic geography perspective (Bull and Church 1994; Urry 1990). The sheer magnitude of the tourism production system for numerous national, regional, and local economies and, more importantly, tourism's powerful interconnections with other economic and social activities occupying the space economy, mean that we can no longer relegate the sector to the periphery of this particular geographic sub-discipline (Townsend 1991).Thus, the principal value of this book rests in its attempt to introduce and legitimize the study of tourism in the eyes of economic geographers and business practitioners.

Prior to laying out the structure of this book, we need to alert our readers to a number of caveats. First and most important, *nowhere is it our intention to argue that tourism must be entrenched solely within the domain of economic geography.* After all, as we pointed out earlier, tourism is far too multifaceted a topic to be isolated to a single sub-discipline of geography. Economic geography is just one of many sub-disciplines within geography that can benefit the study of tourism, particularly as it relates to the sector's more 'industrialized' portions. Specifically, economic and industrial geographers can further contribute towards the development of a more robust conceptual framework of the economics of the travel and tourism production system, as part of a broader theoretical agenda which calls for 'consideration of the commodification of leisure and touristic experiences' (S. Britton 1991: 462). Hence, the scope of this book is far from comprehensive. Instead, through the exploration of the economic geography–tourism nexus, we wish to demonstrate the intellectual value of tourism research within one specific sub-discipline, and concurrently, alert researchers of the virtues of such an approach for a number of other geographic fields.

Second, the book does not provide an exhaustive coverage of all sectors comprising the travel and tourism industry. Instead, it pays attention to certain key sectors and technologies (e.g. airlines, hotels, tour operators, and computerized reservation systems), that are extremely important in terms of their influence on the geography of origin–destination flows.

Third, a final caveat is that the text's geographic coverage is far from universal. Rather, there is a heavy focus on only certain developed capitalist economies. The major case studies presented relate primarily to Australia, Britain, Canada, Italy, New Zealand, and the United States. This geographic bias reflects the book's principal emphasis on tourism's supply-side and the fact that the 'machinery' that facilitates origin–destination flows (e.g. airlines, tour operators, and hotel chains) is headquartered primarily in developed economies. By contrast, examples from less developed and socialist or ex-socialist countries are limited and presented mainly for illustrative purposes.

The book's structure and themes

In sum then, this is a book about the connections of economic geography with the study of tourism. A primary objective is to alert economic and industrial geographers to the intellectual merit of broadening their domain to encompass tourism-related research. Simultaneously, in their efforts to conceptualize the dynamics of the tourism production system, tourism and recreation researchers must be encouraged to make use of the substantial corpus of theories and methods already available to economic geographers.

The book has been written with at least three main audiences in mind. First, it targets tourism specialists in geography but also other disciplines who may be interested in the industry's structure and organization. Second, it aims to serve as a scholarly text for economic geographers interested in enhancing their understanding of the capitalistic nature of travel and tourism, plus these sectors' relationships to other parts of the economy. Finally, this text should be of interest to the various players representing the tourism industry but also policymakers who may wish to enhance their cognizance of the sector's workings.

The book divides into fifteen following chapters organized along five broad themes. Part A comprises three chapters dealing with definitional issues relating to tourism, and particularly the thorny issue as to whether or not tourism can be conceptualized as an industry. Keith Debbage and Peter Daniels (Chapter 2) launch the text by examining the theoretical barriers which, thus far, have impeded the integration of tourism studies within economic geography. They argue that the collection of industries and services which serve the tourist experience (the 'travel industry production system') is an intellectually intriguing economic sector for two reasons. First, it demonstrates a number of 'peculiarities' relative to other more conventional industries, particularly with respect to the 'invisible condition' and the highly integrated nature of production and consumption. Second, it challenges many conventional theoretical perspectives commonly held in economic geography (e.g. economic base theory and the producer/consumer service dichotomy). According to Debbage and Daniels, it is these intellectually challenging issues that make the travel and tourism industry a subject worthy of the attention of industrial and economic geographers.

In the following chapter (Chapter 3), Stephen Smith continues this methodological discussion by arguing that while tourism cannot be thought of as an industry in the traditional sense, there are considerable practical merits to developing definitions which allow this sector to be measured and compared with other parts of the economy. Turning to more practical matters, Smith proposes the tourism satellite account (TSA) as a useful tool for measuring and identifying economic activities which fall within the realm of tourism. Using the example of the Canadian TSA, he demonstrates the advantages of this methodology over the conventional standard industrial classification (SIC) system for classifying and measuring tourism-related activities. Smith argues that despite a number of barriers to developing TSAs, in the long run the benefits far outweigh the disadvantages.

In Chapter 4, Wesley Roehl discusses the definition of tourism-related activities within the realm of traditional industrial classification systems, specifically the standard industrial classification (SIC) system of the United States. He examines the recently implemented North American industry classification system (NAICS) which, as part of the North American Free Trade Agreement (NAFTA), has standardized industrial groupings for Mexico, the United States, and Canada. The NAICS has, according to Roehl, expanded the opportunities for improving the study of travel and tourism-related sectors.

While this book deals primarily with the supply-side of the travel and tourism sector, these issues cannot be treated in isolation from demand-side perspectives. Thus, Part B relates to the determinants of tourism demand and serves as the transition into the main theme of the text. Specifically, Muzaffer Uysal (Chapter 5) overviews the demand-side determinants from a spatial standpoint and discusses their interconnections with tourism's supply-side. He concludes that conventional demand analysis is inadequate for describing and understanding the determinants of tourism demand.

Clear interconnections between tourism and economic geography can be seen in Part C and Part D which focus on the supply-side of tourism. A fundamental purpose of these two sections is to highlight the increasingly complex 'machinery' that governs the production and consumption of tourism, and so, influences the geography of origin–destination tourist flows (see S. Britton 1991: 475). Part C offers a sectoral approach to the travel and tourism industry and focuses upon the key concepts of neo-Fordism and flexible specialization as they relate to the travel industry. These concepts feature as one of the dominant theoretical perspectives of economic geography in recent years.

In Chapter 6, for example, Ioannides and Debbage argue that the travel industry complex is characterized by a polyglot of varying production processes, each placing a premium on flexible forms of accumulation. By examining certain key sectors, they demonstrate how each component of the travel industry polyglot has been affected by flexible-based production strategies such as the externalization of ancillary services, the development of interfirm strategic alliances, and the evolution of sophisticated product differentiation through brand super-segmentation. Ioannides and Debbage end with a call for empirical investigations which will help us to understand further how the travel industry is affected by flexible production techniques and information technologies.

Simon Milne and Kara Gill (Chapter 7) investigate the ability of information technologies (IT), including computer reservation systems (CRS), to transform the travel industry and ultimately develop sustainable tourism products. They critique the existing literature on these topics and stress the need for caution when interpreting the interconnections of the evolution of CRS to development processes in specific destinations. A major finding is that the new information technologies have actually strengthened the degree of industrial concentration in the hands of a few 'giant' industry players. By contrast, small travel-related firms continue to face enormous handicaps in terms of their ability to access CRS and global distribution systems.

The following three chapters examine some of the most important sectors of the travel industry. In Chapter 8, Ioannides analyzes the structural and behavioral characteristics of one of the most under-researched sectors of the travel industry, namely tour operators. He demonstrates how these actors have emerged as powerful manipulators of origin–destination flows, especially in terms of their ability to substitute destinations since they market holiday types and not specific locales. Ioannides ends with a discussion of what the future may hold for this sector. Stephen Wheatcroft (Chapter 9) follows with an overview of the airline industry's impact on tourism evolution in the last thirty or so years. Particularly, he describes the costs of protectionist policies that have traditionally characterized the industry and examines the effects of deregulation and liberalization through case studies from various parts of the world. In Chapter 10, Simon Milne and Corinne Pohlman undertake a case study of the Montreal hotel industry in an effort to enrich our understanding of consumer services, particularly within urban areas. Their analysis focuses on the key challenges the city's hotel industry is currently facing. Milne and Pohlmann maintain that a combination of factors have led to corporate reorganization within the Montreal hotel industry as evidenced by changes in labor management, new technologies, and the development of strategic alliances and networks. Moreover, they argue that hotel chains remain in the best position to survive the competitive nature of the industry.

A discussion of tourism's supply-side cannot be complete without examining the role of the state in developing, promoting, protecting, and regulating a locality's tourism product. Neither can it ignore issues relating to labor and entrepreneurship. Part D, therefore, offers a political economy

11

perspective of tourism, demonstrating how the sector's various stakeholders and institutions interact to develop, promote, and market destinations.

First, Michael Hall (Chapter 11) offers a macro-level perspective of the role of the state in tourism, an issue that has been very much neglected in the past. Specifically he discusses the role and function of governmental institutions in tourism. He stresses that the conflicting priorities of the various public sectors with an interest in tourism are largely caused by the fragmented nature of tourism itself. The end result is non-coordination and thus, ineffective state control of the industry. Hall also argues that similar to other components of the space economy, tourism has been affected by recent changes in the political atmosphere of western economies, particularly the moves towards deregulation and privatization as well as the globalization of capital. This has meant that in this era of 'smaller' government, places increasingly have to fight for their economic survival on a global level. One manifestation of these 'place wars' is the emphasis localities have placed on attracting tourists.

This scenario has become the established norm in cities throughout the United States as Robert Beauregard points out in Chapter 12. He argues that, whereas forty years ago most localities neglected tourism in their economic development plans, in this age of economic restructuring the industry has gained enhanced respect as a tool for economic growth. According to Beauregard, among the factors which have led to tourism's elevated status in the United States are lifestyle changes, the need for city governments to find new sources of revenue, and a transformation of the industry itself. He ends with a dire warning that while tourism is currently in the limelight as a priority sector for urban economic development, it is vulnerable to a number of factors which could push the industry back to the periphery.

The following two chapters turn their attention to issues of entrepreneurship and labor within the tourism sector. In Chapter 13, Gareth Shaw and Allan Williams argue that a major reason behind our limited understanding of tourism's effects on economic change arises from the failure of most tourism geographers to consider entrepreneurial activity within the sector. They begin with an overview of the concept of entrepreneurship in the tourism sector, drawing primarily from studies relating to less developed countries. Moreover, they discuss the need to incorporate detailed perspectives of entrepreneurial activity within evolutionary models of tourism development (e.g. the resort life-cycle model). Shaw and Williams point out that, although spatial and temporal contingencies dictate the nature of entrepreneurship in different destinations, these issues have thus far not been thoroughly researched. In the latter part of their chapter, they report on the state of entrepreneurship in the British hotel industry, paying particular attention to the characteristics of the owners of small accommodation establishments in coastal resorts.

Understanding the characteristics of tourism-related labor is also an important theme that ties the study of this sector to economic geography. Unfortunately, most existing geographical studies on tourism have rarely offered in-depth perspectives of labor-related issues in the travel and tourism sector. Paul Mackun (Chapter 14) examines entrepreneurship, labor characteristics, and the role of the state through a case study of tourism development in the Province of Rimini, located in what has become commonly known as the Third Italy. As he points out, in recent years the Third Italy has attracted the attention of a number of economic geographers interested in the effects of flexible means of production on small and medium-sized manufacturing firms. Drawing analogies from this literature, Mackun discusses patterns of tourism development in the Province of Rimini. Among other issues, he pays close attention to patterns of ownership among tourism establishments, laborers' characteristics, the evolution of business networks among tourism employers, and government policies as they relate to the travel and tourism sector.

Longitudinal studies relating to the behavior of various sectors are yet another important research area in economic geography. Nevertheless, the effects of economic business cycles on the evolution of travel-related sectors have been largely ignored. Significantly, there has been very little work on the effects of innovation on the trajectory of tourism development in various destinations. In Chapter 15 (Part E), Michael Haywood examines the sensitivity of the demand and supply-side of travel-related sectors to economic business cycles. Following a detailed critique of the life-cycle model of destination evolution, he focuses on a case study of the lodging industry to illustrate the travel sector's sensitivities to national as well as global recessions or expansions. Finally, Haywood notes that tourism, like many industries, is affected not only by swings associated with business cycles but also by long waves of economic development (e.g. technological innovations and entrepreneurial discoveries). Indeed, according to Haywood, product innovations have significantly accelerated the life cycles of travel-related organizations (e.g. hotel chains).

In Chapter 16 (Part F) the concluding part of this edited collection, Debbage and Ioannides revisit the key issues discussed, once more underlining the significance of incorporating studies of tourism's supply-side within the mainstream of economic geography. The final pages of the book offer an agenda of new research directions which will help enrich the economic geography of travel and tourism.

Part A

CONCEPTUAL AND DEFINITIONAL ISSUES: BARRIERS TO THEORY

2

THE TOURIST INDUSTRY AND ECONOMIC GEOGRAPHY

Missed opportunities?

Keith G. Debbage and Peter Daniels

> Tourism has been relatively neglected by the social and economic sciences, particularly in light of its economic importance. There are a number of reasons for this neglect. First, and perhaps most significant, scholars have been reluctant to turn to leisure studies, they have traditionally been more attuned to research on work and 'hard' production. Activities that produce enjoyment, even fun, have been disdained, either because the research itself might be pleasant or because only 'serious' subjects merit attention. It all seems a bit puritan.
>
> (R. Britton 1978: 17)

The tourism sector: an invisible industry?

The travel and tourism industry is an increasingly important part of the employment structures of advanced industrial nations and lesser developed countries. Many of the tourist industry's key suppliers are headquartered in the developed world (e.g. American Airlines and Holiday Inn), while tourist expenditures tend to be disproportionately significant for third world economies, especially in the Caribbean and South Pacific. The tourist industry has become a highly dynamic spatial network of production and consumption, and as such is implicated in some of the critical theoretical issues of current concern to economic geographers: the globalization of capital and firms (Berry *et al.* 1993; Dicken 1992); deindustrialization and regional economic restructuring (Goe and Shanahan 1990; Harrison and Bluestone 1988); the increased significance of strategic alliance networks (Debbage 1994); the spatial division of labor (Massey 1984; Walker 1985); urban revitalization (Mullins 1991); the growth of the information technology services-based economy (Hepworth 1989); the evolution of advanced services especially producer services (Beyers 1992; Daniels 1986; Drennan 1992;

17

Greenfield 1966); and the creation of post-modern/post-industrial/post-Fordist landscapes (Harvey 1989a; Urry 1990 and 1995).

Despite this, not much has changed since R. Britton's provocative comments in 1978. Although the tourist industry, along with other service industries such as retailing or financial services, has greatly expanded its role in many national economies, the industry continues to be a grossly misunderstood and underexplored sector of the world economy. Consumer services like tourism are considered by many to be 'parasitic' and biased towards low-paid, female and part-time employment, such that tourist jobs are not considered to be 'real jobs'. It is a conviction that persists among many economic geographers who are particularly guilty of ignoring or under-playing the important role that tourism plays in shaping the spatio-economic landscape. It is fairly common for current standard textbooks of economic geography largely to ignore the role that tourism plays in the global, regional, and local economy (Berry *et al*. 1993; de Souza 1990; Dicken 1992). All this is so, even though the tourist industry as a mode of production is enormous, highly commodified, and structured in ways that are fairly similar to other sectors of the economy.

While it is still fairly easy to find geographers who are unable to contemplate how anything but extractive industries and manufacturing can form the economic base of a regional economy, the tide appears to have turned. For example, several geographers (S. Britton 1991; Debbage 1992; Ioannides 1995 and 1996a; Townsend 1991) have recently called for a strengthening of the ties between tourism geography and the mainstream economic geography literature. According to Townsend (1991: 36), 'we are long past the day when factory production could be regarded as closely congruous with the "economic geography" of a typical provincial area'. Furthermore, in a recent progress report on industrial geography, Malmberg (1994: 532) argued that industrial geography should encompass 'not only the location of manufacturing activities but also the wider processes of struc-tural change in the production of goods *and* services'. The rationale for such a position is partly related to the increasingly blurred distinctions that now exist between many forms of manufacturing, the services, and the producer service economy, and the increased role of the latter within the world economy. According to Daniels (1993: 2), the 'service industries now account for at least two-thirds of total employment in developed economies and for at least 50 per cent of gross domestic product'. As a result, the study of the magnitude and role of service-sector economic activities has become a significant research theme in economic geography over the past decade (Beyers 1992; Daniels 1993; Marshall *et al*. 1988).

This chapter explores some of the theoretical challenges that arise when attempting to integrate one of the most important consumer services in the world – tourism – within the established practices and literature of economic and industrial geography. The primary mission is to illustrate that the many and diverse components of the travel industry production system can be integrated into the broader themes of economic geography. By doing so, we hope that this chapter can enrich the existing theoretical literature on the geography of production and consumption while, at the same time, enhancing the legitimacy of the study of tourism and the tourist industry as an organized system of commodity production, not unlike any other sector of the economy.

We begin with an overview of the definitional and classification issues relating to services and tourism, and briefly articulate the position of tourism within the producer/consumer service dichotomy. Next, we review the loose language and the confusing terminology commonly used when defining what comprises the travel and tourism industry, and examine some of the 'peculiarities' of the travel production system. The chapter concludes with an overview of economic base theory as a traditional explanation of economic growth in the service industries. We discuss the difficulties that economic base theory has posed for a tourist industry that does not neatly fit into the basic/non-basic categorization of an economy.

Services definition and classification

According to Begg (1993: 817), 'the problem of how best to define the service sector has filled many column inches of print' and no universally accepted categorization has emerged (Daniels 1985; Fuchs 1968; Gershuny and Miles 1983; Marshall *et al.* 1988; Stanback 1979; Urry 1987). Part of the conceptual difficulty can be related to the intangible nature of many service-based economic activities, and the limited shelf-life of many service-based products. Most services do not involve the sorts of tangible physical transformations that we see in goods-producing sectors like the automobile industry. Furthermore, according to Daniels (1993), some services yield only immediate or short-term utility such as the so-called consumer services (e.g. entertainment-based activities like a movie theatre or a restaurant). These consumer services are largely based on a set of labor inputs where production and consumption are simultaneous and the end result is often simply an experience, rather than a tangible good. Other services offer a more extended or semi-durable form of utility and these would include activities such as automobile repair, taxation advice, and dental treatment. Still other services are even more durable and provide much longer-term utility (e.g. mortgage financing, life insurance, and pension arrangements).

Although these examples illustrate the elusiveness of services, it is also important to remember that tangible manufactured goods, like an airplane, also provide services to the consumer. The 'value' of this service will depend on how it is used. For example, an airplane can provide a wide range of different services based on: the proportion of seating allocated to business class passengers rather than leisure class passengers, the amount of leg-room between seats, the type of in-flight service provided, the volume of freight carried in the underbelly, the ability of the airline owner to properly maintain the aircraft, and the type of city-pair routes flown by the air carrier. This blurring of goods-production and service-providing activities has led some to attempt to group services into those meeting final demand (i.e. the consumer services) and those meeting intermediate demand (i.e. the producer services) (Marshall *et al.* 1988).

> The basic distinction between consumer and producer services needs little elaboration. Those services that at destination are used by households or individuals fall clearly under the former, and those services used ultimately by business firms and other productive enterprises are included in the latter.
>
> (Greenfield 1966: 7–8)

To this end, much attention has been paid to the producer service sector, which has been said to possess many of the attributes ascribed to manufacturing (Daniels and Moulaert 1991; Illeris 1989; Marshall *et al.* 1988). Producer services have evolved in response to intermediate demand because they are those services which supply other businesses and government organizations which in turn provide additional service-based outputs and products. They can be grouped into those services focused on information processing activities (e.g. banking, insurance, marketing, accounting, advertising, software firms); goods-related services (e.g. distribution, transportation, maintenance); and finally, personal support services (e.g. catering, personal travel, and accommodation) (Daniels 1993; Marshall *et al.* 1987). According to Daniels (1993), it is precisely these sorts of services that can play a key role in enhancing the efficiencies of various production processes, in stimulating technological innovation, and in maintaining the competitive advantage of regional economies. We should note here that it is not too much of a stretch to consider tourism as a type of personal support service if it is considered to include catering (e.g. restaurants), personal travel (e.g. vacation time), and accommodation (e.g. the hotel industry), especially if the services provided cater to the business travel market – a point to which we shall shortly return.

Despite this intense academic interest in the producer services, others have argued that considerably less energy has been expended on investigating the role of the consumer service sector in satisfying final demand (that is, services which supply private individuals or consumers) (Townsend 1991; Williams 1996). Part of the reason why the consumer services have been relatively neglected is that they are considered theoretically moribund. They conform to a static Christallerian vision of central place location where services essentially chase those higher-income markets that are most likely to induce spending on consumer-based products. In this sense, consumer services are viewed as not tradable since they typically can only be directly consumed or purchased at 'fixed' points of supply (e.g. clothes retailing and fast-food restaurants). Consumer services like these are considered 'parasitic' since they feed-off the wealth generated from other sectors through such things as disposable income, and serve only the local market because they require physical proximity to the purchaser. It will be suggested later in the chapter that this viewpoint is an oversimplification.

Although the producer service/consumer service dichotomy has been widely adopted, a good deal of concern remains about its conceptual accuracy as the base of a classification system for services (Allen 1988; Walker 1985). According to Daniels (1993: 4), 'the so-called consumer and producer services are not mutually exclusive groups; some service activities such as banking, insurance, and finance fulfill both final and intermediate demand'. Bull and Church (1994) raise similar concerns in an analysis of the British hotel and catering industry. They argue that the hotel and catering industry simultaneously behaves as a consumer service when responding to local demands and as a producer service when catering to business demands. For example, a hotel functions as a consumer service because it offers drinking and eating facilities frequented by local inhabitants. By contrast, Bull and Church (1994: 21) argue that 'the marked growth of business travel is the most obvious manifestation of producer service demand for hotel and catering facilities'. Daniels (1986) estimated that approximately 25 per cent of all hotel and catering employment in Great Britain is producer-service based.

The emergence of a post-industrial or information-intensive society (Bell 1973; Hepworth 1987 and 1989) has increased the volume and importance of information exchange and the development of business contacts, resulting in a rapid growth in business conferences and business-related travel – a form of intermediate demand. According to G. Smith (1990), approximately 62 per cent of conference-related expenditure in the United States is accrued by the hotel industry, while Zelinsky (1994) found that United States convention goers had increased from 5.9 million in 1965 to 13 million by 1990. This has prompted Randall and Warf (1996) to analyze the economic impact of the annual conferences of the Association of American Geographers on the host cities, as it becomes more widely accepted that parts of the travel industry satisfy intermediate as well as final demand.

One solution that can help minimize some of the limitations of the consumer/producer service dichotomy is to estimate the number of employees contributing to final and/or intermediate demand. As noted by Stephen Smith in Chapter 3, the construction of tourism satellite accounts may be one way effectively to disentangle final and intermediate demand in the travel industry. The problem is not, however, simply one of assigning elements of the tourism production system to intermediate and final demand; some critics even question the existence of a tourist industry at all.

The tourism–tourist–travel conundrum

If, as Daniels (1993: 1) argues, the 'service industries have long been the Cinderella of economic geography' then it surely must be the case that tourism and the travel industry have been the 'Cinderella' of services (Ioannides 1995). According to Mullins (1991: 326), 'although tourism is considered the fastest growing industry in the world today, it is a little understood industry'.

Part of the problem rests with the amorphous nature of tourism and the conceptual and methodological complexity underlying terms like tourism, tourist, and travel. The lack of consensus on what these terms mean has resulted in a certain level of 'conceptual fuzziness' that has triggered an on-going debate between geographers, and other social scientists, about the validity of treating tourism as an industry. In this debate, tourism is typically conceived of as any good or service that has been consumed by a tourist, and herein lies the methodological conundrum. Rather than focusing on defining the travel industry as a provider of goods and services, or as a system of production which facilitates access to fixed assets such as landscapes or museums, much of the literature attempts to define the industry based on definitions of a tourist as a consumer.

> Goods and services are classified as 'tourist' if they are consumed by tourists, but if they are consumed by residents they are considered part of 'normal' consumption. Largely for this reason, tourism research has become a form of market research, with official statistics, for example, being collected almost exclusively on consumption – the demand side – and very little being collected on the production and distribution of goods and services – the supply side.
>
> (Mullins 1991: 326)

S. Smith (1988: 183) viewed such an approach as equivalent 'to defining the health care industry by defining a sick person'. Such a state of affairs has made it extraordinarily difficult to pinpoint both the fundamental nature of tourism and the economic forces that have shaped this industrial sector – a point to which we shall return later in this chapter.

From a technical viewpoint, the individual tourist is conceived as any person completing at least a one-night stay away from the place of permanent residence (although that stay must not exceed one year and may not involve remuneration following an occupation). Institutions such as the Statistical Commission of the International Union of Official Tourist Organizations in the 1960s and, more recently, the World Tourist Organization have expended considerable energy to ensure that this definition is followed so that the data collected are comparable across countries.

While the explicitly spatial nature of such a definition has been of great interest to geographers, it has effectively defined tourism negatively with respect to the normal place of residence. Tourism becomes a residual 'catch-all' that includes literally any activity that takes place away from the home provided it occurs within the time frame of twenty-four hours to one year (G. Hughes 1991). By defining a tourist, and thus tourism, in this way, it becomes possible to embrace an enormous diversity of trip purposes provided they are 'performed "away" rather than at "home"' (G. Hughes 1991: 265). Consequently, the tourist industry can involve an individual on vacation but it could also be a conventioneer, a shopper, funeral and wedding guests, or an individual visiting friends or relatives. Such an open-ended definition has made it easy for 'boosters' to over-inflate grossly and hype the industry as the 'largest in the world'.

Although some aspects of tourism are exaggerated, it is an increasingly significant industry that desperately requires a more rigorous conceptualization of what this thing tourism really is. According to Mullins (1991: 326), a more useful approach may be to focus, not on the tourist, but on the industry itself, 'although even here there are difficulties because tourism is a very odd industry'.

Certainly, tourism is a fundamentally different type of industry from other forms of commodity production. Tourism is no single product but, rather, a wide range of products and services that interact to provide an opportunity to fulfill a tourist experience that comprises both tangible parts (e.g. hotel, restaurant, or air carrier) and intangible parts (e.g. sunset, scenery, mood). For economic geographers, this means that the tourist industry poses many unusual development implications. According to Mathieson and Wall (1982), these 'oddities' include the following:

1   Consumers visit the site of production (or 'factory'), rather than the good being transported to the user. This means that tourism can be described as an 'invisible industry'.
2   Expenditure can be substantial (e.g. an airfare or hotel accommodation) and the product is often purchased unseen.
3   There is often no tangible return on investment because the tourist product is primarily an experience, not a good.
4   Purchases are rarely spontaneous but carefully planned (especially in terms of destination choice, accommodation type, and mode of travel).
5   The product is highly perishable and cannot be stored and sold at a later date (e.g. if a hotel bed or airline seat is not occupied, then that rental opportunity is lost and cannot be replaced).
6   Distance cannot always be regarded as a disutility in commodity transactions because not all tourists are distance minimizers (e.g. some tourists are interested in visiting new, more distant and exotic destinations, despite the additional costs).

While many other sectors of the economy may share some of tourism's unique attributes (e.g. offshore banking is an 'invisible industry' and the poultry industry is highly perishable), it is difficult to find a sector with such a challenging array of complex and unusual forms of commodification. In this sense, tourism can be viewed as an intellectually challenging avenue of research, particularly in the context of the post-modern/post-industrial/post-Fordist era, and the increased emphasis placed on better understanding the commodification of culture and the so-called 'tourist gaze' (S. Britton 1991; Featherstone 1991; Harvey 1989a; Urry 1990). Indeed, the actual purchase and consumption/production of tourist services (the airline seat, the meal, the admission ticket) may actually be incidental to Urry's well-worn concept of the 'tourist gaze', defined here as the time spent gazing upon such objects as a work of art, a historic attraction, or a scenic vista. Furthermore, although the actual time spent participating in a 'tourist gaze' may be extremely brief, it can be central to the overall quality of the tourist experience and, of course, is inherently spatial since it implies a movement of people to some distant destination.

To date, most of the geographical models of tourism have failed to incorporate the travel industry – as a supplier of goods and services that help to facilitate the tourist experience or gaze – into the spatial context (Stabler 1991). A similar point has been made by S. Britton (1991: 451) who claims that 'geographers working in the field [of tourism] have been reluctant to recognize explicitly the capitalistic nature of the phenomenon' even though it is the travel industry (e.g. airlines, travel agents, tour operators) that can most influence the geography of origin–destination tourist flows. For example, while Mathieson and Wall (1982) make a persuasive case that tourism has a number of unique attributes, all the 'oddities' listed earlier focus on the end product or tourist experience, rather than on the infrastructure of production.

This book should be viewed as a first attempt at correcting this deficiency – although the debate about the legitimacy of tourism and travel as an industry is a long-standing one. In geography, the debate has been stimulated by a series of commentaries in the *Canadian Geographer* in the late 1970s and early 1980s (R. Britton 1979 and 1981; Chadwick 1981; Mieczkowski 1981), and more recently, in a combative exchange between S. Smith (1988, 1991, 1993 and 1994) and Leiper (1990a and 1993b) in the *Annals of Tourism Research*.

Part of the debate in *The Canadian Geographer* focused attention on the 'conceptual fuzziness' that prevails when using the terms tourism and travel. Mieczkowski (1981) argued that the tourism industry is commonly referred to as the travel industry because of the negative connotations of some forms of mass tourism (e.g. low average expenditures per visit and the 'Ugly American' caricature). Underlying much of the debate was the argument that tourism may not be a coherent industry since it does not produce a distinct

Unless there is a renaissance in manufacturing employment and an improvement in the wages offered by smaller subcontractors or a substantial upgrading of jobs in the service and trade economy, the deindustrialization hypothesis predicts that the stagnation in real wages, growth in the low-wage share, and polarization of the entire earnings distribution will continue – regardless of expected demographic trends and the state of the business cycle.

(Harrison and Bluestone 1988: xvi)

The rationale for this perception, in part, has its roots in 'economic base' theory which is built on the assumption that an area needs to generate export sales in order to grow in a healthy fashion. Under this analytical framework, any economy is divided into either a basic or non-basic sector. The 'basic' industries have traditionally included extractive industries and manufacturing firms that are capable of bringing export-based revenue into the area, while the service sector activities are seen as 'non-basic' or dependent industries that serve only the local market and offer low-wage opportunities. Under such a theoretical construct, the contention is that services are wholly dependent on the wealth generated by the high-wage manufacturing sector (in terms of disposable income) and that service jobs are not 'real' jobs (the so-called 'hamburger flipperization' or 'Walmartization' of the job market).

However, during the past decade, the growing prominence of service activities has led to some research workers questioning the simplistic assumptions of economic base theory (Begg 1993; Goe and Shanahan 1990; Marshall and Wood 1992; C. Williams 1996). According to Urry (1987: 6), 'it is incorrect to view services as essentially "non-basic" by contrast with supposedly "basic" manufacturing'. Urry (1987), and others (Daniels 1993; Goe and Shanahan 1990) have argued that the service industries are now too diverse to be adequately captured in a crude basic–non-basic dichotomy. Such a dichotomy, they argue, could not possibly adequately capture the heterogeneous nature of many service sector activities, that frequently possess fundamentally different economic characteristics. Furthermore, the markets for many services now extend beyond the regions in which they are produced, and some services can even bring external income back into the area. For example, the accelerating demand for high-skill producer services provides a partial explanation for some of the growth in high-wage service jobs. This is because many of these products satisfy intermediate demand and are in some sense tradable since they do not necessarily require a physical proximity to a purchaser (e.g. information processing services).

An additional conceptual problem is that many services are not physically transferrable out of the region in which they are produced, even though they may have a capacity for generating externally-derived income for the area. This is the case with many tourist-related facilities that can only be consumed by individuals visiting the site of production even though they may have traveled a considerable distance to consume that product. Partly because of these conceptual difficulties, C. Williams (1996) has argued that the basic sector should now be recognized as comprising not only industries which export tangible goods or services (e.g. manufacturing and the producer services), but also industries which generate external income through the provision of less tangible services (e.g. basic consumer services such as tourism, regional malls, and sports stadiums). However, classifying tourism as a basic consumer service may be partly misleading since others have argued (Begg 1993; Bull and Church 1994) that some forms of the industry, such as business tourism, can be sold to other firms as a producer service.

Additionally, while some have argued that the producer services are a key factor in perpetuating uneven development within space economies because their demand and supply need not be geographically coincident (Marshall *et al.* 1988), it is clear that the demand for tourist services may be equally complex and as spatially diverse. Although the production and consumption of the tourist experience tend to occur simultaneously, they rarely occur at *only one fixed point of supply* because of the enormous complexity involved in any given individual origin–destination tourist flow. An individual is consuming a tourist experience immediately upon departure from his or her household, and the 'machinery' involved in the production of that experience is sophisticated, geographically diverse and spatially uneven (e.g. airline, hotel, restaurant, souvenir store, cruise ship, travel agent, tour operator, car rental operator, casino, theme park, etc.). The spatial variation in the factor endowments for tourist services (natural resource attractions such as a beach or mountain; human-built attractions such as Disney World or a casino; and cultural attractions such as a historic building or battlefield) are highly uneven between countries, regions, and cities. Furthermore, although the provision of tourist services is to a significant extent spatially fixed, the tourist is increasingly mobile and able to consume tourist services on a global basis. For these reasons, tourism may be just as likely to perpetuate uneven development within space economies as are the advanced producer services (Townsend 1991). Furthermore, 'an important factor in the growth of producer services has been the progressive out-sourcing of demand' (Begg 1993: 818) and there is considerable evidence that this is occurring in the travel and tourism industry (see Chapter 6 by Ioannides and Debbage). This phenomenon should not be surprising since this is a system of production which requires specialist inputs such as advertising, inclusive tour charters, and financial services, so that the component parts of the tourist industry can remain competitive.

Indeed, many of the key issues facing the diffuse tourism sector are not uncommon to the manufacturing sector or the producer services. These issues include: the elevated significance of information technologies and computer reservation systems (e.g. automated check-in/check-out at the major hotel chains); the increased importance of sub-contracting and numerical/functional

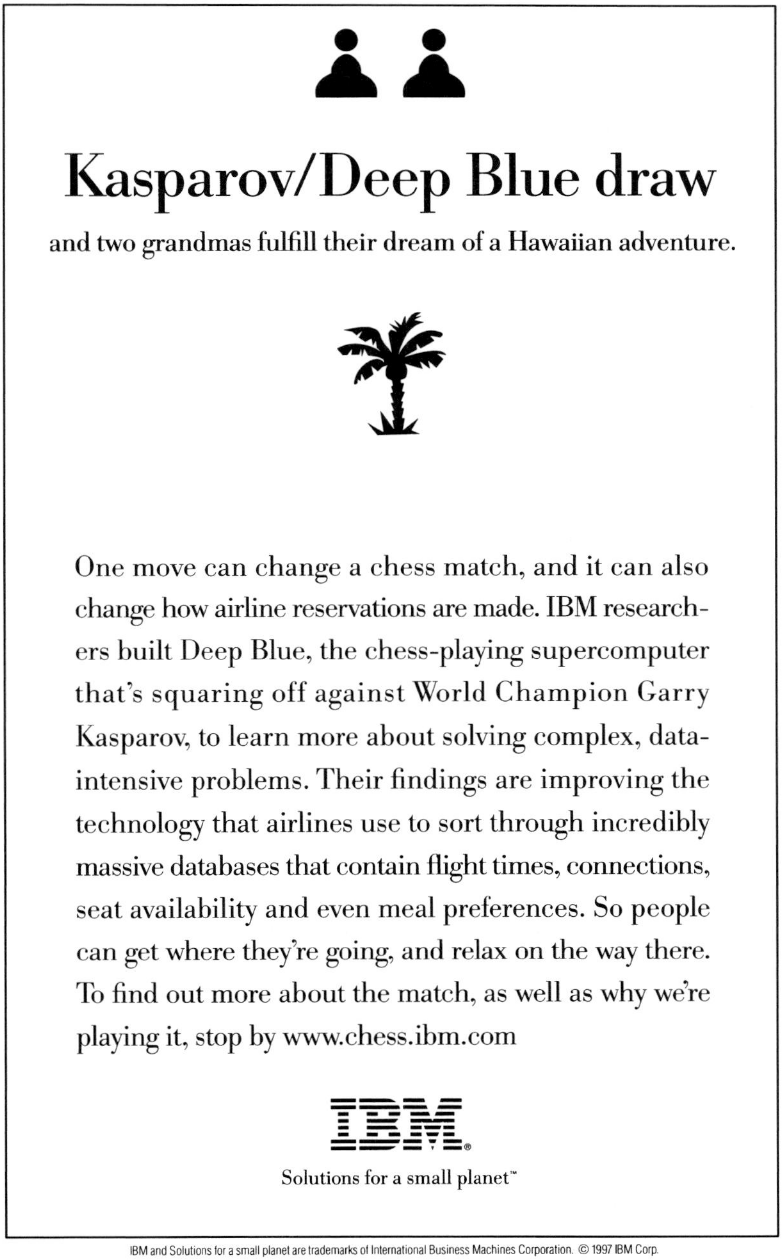
Kasparov/Deep Blue draw
and two grandmas fulfill their dream of a Hawaiian adventure.
One move can change a chess match, and it can also change how airline reservations are made. IBM researchers built Deep Blue, the chess-playing supercomputer that's squaring off against World Champion Garry Kasparov, to learn more about solving complex, data-intensive problems. Their findings are improving the technology that airlines use to sort through incredibly massive databases that contain flight times, connections, seat availability and even meal preferences. So people can get where they're going, and relax on the way there. To find out more about the match, as well as why we're playing it, stop by www.chess.ibm.com
IBM
Solutions for a small planet™
IBM and Solutions for a small planet are trademarks of International Business Machines Corporation. © 1997 IBM Corp.

Figure 2.1 Travel-based advertisement for IBM's chess-playing supercomputer
Source: *The New York Times*, 6 May 1997

flexibility in the workforce; the spatial separation of front-office and back-office functions (e.g. the data processing of American Airlines ticket information from Barbados); and the commodification of culture through specialized tourist-based 'consumption compounds' like Boston's Fanueil Hall or Baltimore's Inner Harbor (Mullins 1991).

Conclusion

We hope that this chapter has cast some light on the complex problem of how to integrate the tourism geography literature with the much larger literature on the geography of production and consumption. Part of the methodological conundrum facing tourism studies is effectively disentangling the consumer from the production system and rigorously conceptualizing what this thing tourism really is. The theoretical difficulties posed by such a problem should not be used, however, as a reason to continue to ignore this critically important part of the space-economy. In fact, the travel production system is, in many ways, one of the most intellectually intriguing sectors of the economy, because it challenges the conventional paradigms with regard to economic base theory and the consumer/producer conceptualization of services.

Economic and industrial geographers who have focused their attention on service activities seem to have begun to recognize this very fact in their recent calls for the development of a more ambitious conceptual framework that can illuminate the nature and role of *all* services, whether they be consumer, producer, or public-sector oriented (Dunham 1997; Marshall and Wood 1992; Townsend 1991). We believe that there *is* a need for a more holistic research agenda that expands our intellectual horizons to include a more rigorous examination of what may be the largest service-based sector in the world.

While it is not the purpose of this chapter to articulate a lengthy research agenda for economic geographers interested in tourism, we would suggest that a recent advertisement carried by IBM (Figure 2.1) illustrates the potential significance of future research in this field. Such is the spatial complexity of origin–destination tourist flows by air that IBM is utilizing 'Deep Blue' – the chess-playing supercomputer – to help to improve the way in which airlines route travelers from point A to point B. We believe that this is the intellectual tip of a very large, and largely underexplored, iceberg.

3

TOURISM AS AN INDUSTRY

Debates and concepts

Stephen L. J. Smith

Conceptualizing the supply-side of tourism

The phrase 'tourism industry' is used so often that one can begin to believe it actually means something. This chapter examines what that meaning might be. More precisely, this chapter examines the issues and debates surrounding two questions: Is tourism an industry? How might one measure the magnitude of tourism as an economic activity, whether or not it is an industry?

The significance of the question of whether tourism is an industry can be seen in the following examples. Tucker and Sundberg (1988), in a review of international service trade, summarized (and virtually dismissed from further discussion) tourism with the observation:

> It [tourism] is not an 'industry' in the conventional sense as there is no single production process, no homogeneous product, and no locationally confined market.
>
> (Tucker and Sundberg 1988: 145)

The implication of such an observation is that the tourism 'industry' does not need any special analysis or policy consideration *per se* because it simply does not exist. Tourism-related sectors such as transportation or accommodation may be worthy of policy consideration because they are industries 'in the conventional sense', but tourism as a whole is not.

In contrast, S. Smith (1988) proposed what he called a supply-side definition of tourism:

> Tourism is the aggregate of all businesses that directly provide goods and services to facilitate business, pleasure, and leisure activities away from the home environment.
>
> (S. Smith 1988: 183)

The implication of a supply-side definition is that such a definition would be fundamental to any consideration of tourism as an industry. As Tucker and Sundberg (1988) imply, an industry is defined in terms of a homogeneous product and single production process.

Leiper (1990a) countered that, while Smith positioned his definition as a supply-side one, data required to measure tourism must come from demand-side sources, and thus – operationally – Smith's definition is not supply-side. Leiper further argues against Smith's definition with the following analogy:

> Saying that a firm is in the tourism business or industry merely because it has customers who can be described as tourists is . . . analogous to observing red-heads among the customers of the butcher, baker, and candlestick maker and deducing the existence of a 'red-heads industry'.
>
> (Leiper 1990a: 602)

Leiper goes on to suggest that tourism might be best described as 'partially industrialized'. By this Leiper seems to mean that visitors are served, as part of a tourism experience, both by firms that cater almost exclusively to visitors as well as by other firms for whom visitors are only a small part of their customer base. Smith and Leiper continued their debate in S. Smith (1991), Leiper (1993b), and S. Smith (1993). The positions are, in a nutshell: Smith asserts that, while tourism is not a conventional industry, it can be defined and measured in a way that is consistent with other industries; Leiper counters that tourism is best viewed as a mix of industries that have varying involvement in the provision of services to visitors, and should not be characterized as a single industry, conventional or otherwise.

Besides Smith and Leiper, who cares? The expression 'tourism industry' is most commonly used in an advocacy context. The term is a convenient way of referring to the large and fragmented collection of firms producing commodities that support the activities of people temporarily away from their usual environment. These commodities include transportation, accommodation, food services, attractions, and travel trade services. Although these businesses traditionally do not cooperate much with each other to promote their common interests, they sometimes rally together to express frustration over government policies, regulations, and taxation systems that affect them all. They sometimes will even form organizations with names like 'Tourism Industry Association of X', where X is a nation, province, or state. One message that such organizations try to send to policymakers is that governmental actions not only affect individual businesses, but are also something of strategic importance in the regional or national economy – the 'tourism industry'. Estimates are then made of the industry's contribution to GDP, employment, payrolls, and government revenues. The implicit belief is that thousands, or tens of thousands, of primarily small and medium-sized enterprises scattered over maybe a couple of dozen different sectors will have greater political clout if they can present themselves as one large industry: a 'strength in numbers' strategy.

This strategy is a plausible one. Everything else being equal, a large industry supporting many workers and creating large amounts of export earnings and governmental revenues is more likely to influence government decisions than a small industry. In the case of tourism, however, the implementation of this strategy has not always been successful. In many nations, the industry does not really view itself as an industry – which means government is not likely to view tourism as an industry either. Firms involved in tourism tend to view each other as competitors rather than as potential partners or allies. If they do form partnerships, the partnerships tend to be either highly specific, such as a resorts association or a campground owners' association, or are geographically-based, such as a destination marketing organization (DMO). The existence of either a sectoral association or a DMO is not evidence that tourism is being viewed as an industry (whether by government or the members of the organization). Rather, these are *ad hoc* organizations created simply to do some joint marketing or lobbying. In fact, it is not uncommon to find DMOs and sectoral associations striving against each other for memberships and political influence. It is rare to find integrated, industry-wide, cooperative marketing strategies, with a commitment to sharing data and research, and a willingness to work together on industry-wide challenges.

One of the most important reasons for such fragmentation has already been implied: the great majority of most nations' tourism businesses are small or medium-sized enterprises. Small firms are usually more concerned with day-to-day operations than with strategic partnerships and industrial strategies. Time and money are too limited to spend either on organizations or activities that do not directly help them with their immediate needs. Further, owners and managers of tourism firms often do not have formal business training. The operation may be a family business started by an individual who had the entrepreneurial skills needed to create a business, but not necessarily the political intuition or experience, or an interest in forming strategic alliances to promote industry interests.

Another barrier to the acceptance of tourism as an industry is the long-standing problem of a lack of credibility in tourism statistics. This is especially relevant from the governmental perspective. The lack of credibility stems from a number of roots. One of the more important is inconsistency in definitions and units of measurement. One organization defines tourism as exclusively pleasure travel, another includes visiting friends or relatives (VFR), yet another includes both of these plus business travel. Different jurisdictions may define a 'trip' for the purposes of counting visitors and their expenditures in terms of a minimum length of stay; minimum distance travelled; or whether an international border was crossed.

Tourism employment may be measured in units of 'full-time equivalents' (a job held by a person working full-time for one year, or the equivalent such as two people working full-time for six months); a 'person-year' (a person employed for twelve months regardless of the hours worked per week); 'positions' (a job description in a firm); or simply the number of actual human beings working during a specified period in tourism organizations, regardless of the number of hours or whether the position is permanent or temporary. Each of these measures has utility, but they lead to different estimates and conclusions and, ultimately, to the impression that no one in tourism knows what they are talking about.

Credibility problems in tourism data also arise from the quality of data sources. Sample sizes, period of recall (i.e. whether the person is asked questions about a trip just completed, one taken three months ago, or one taken a year ago), the wording of questions, and other methodological details affect the precision, accuracy, validity, and reliability of tourism statistics and, by association, the industry's credibility.

An example of the importance of definitional conventions and the confusion resulting from violating them can be seen in claims made by the World Travel and Tourism Council (WTTC 1996b). By convention, industries are defined in terms of final products. For example, the fisheries industry consists of those businesses that produce fish; the steel industry is the set of firms that produce steel. The tourism industry is those firms that produce tourism commodities consumed by visitors in the course of a tourism experience. The WTTC, in contrast, claims not only the output of such firms as part of what they call 'travel and tourism' (inconsistency in terminology is yet another source of doubt about the credibility of tourism for people outside tourism), but also certain shares of the output of construction and manufacturing industries. While tourism firms generate demand for the products of these industries, the output of the construction and manufacturing industries cannot meaningfully be claimed as tourism commodities. Doing so is comparable to the motor vehicle industry, for example, claiming a portion of the output of mining as part of motor vehicle production because motor vehicles create a demand for metals.

The impact of this non-conventional definition of tourism can be appreciated by looking again at the example of Canada. The WTTC estimates that 'travel and tourism' account for about 13 per cent of Canadian GDP in 1988 (WTTC 1995). Statistics Canada, measuring the output of the 'tourism industry' in a way comparable to other industries, estimates the contribution to GDP at a mere 2.5 per cent (Lapierre and Hayes 1994). The WTTC methodology inflates the importance of tourism in Canada by more than five times. (A comparison of the methods of the WTTC and Statistics Canada can be found in Wilton (1996)). While WTTC's goal of increasing the political clout of tourism is understandable, their extravagant estimates are not defensible and undermine the industry's credibility. As an editorial in *The Economist* (1995) observed:

> The WTTC, a Brussels-based lobbying group, claims that tourism is Europe's biggest industry. . . . To lump together airlines, farmers who supply vegetables to resort towns, and the construction companies that build hotels is statistical sophistry. . . . [To] use this as an argument for subsidy is special-interest lobbying at its most naked.
>
> (*The Economist* 1995: 13)

Ultimately, the discrepancy between Statistics Canada and the WTTC results not from differences in definitions or methods, but in their core values. Statistics Canada, as a national statistical body, is concerned with ensuring credibility, consistency, objectivity, and comparability in all its estimates of the outputs of industries. It thus collects and analyzes statistics for all industries in a consistent, comparable fashion. The WTTC, as *The Economist* noted, is an international lobby organization concerned with generating the largest number possible to support its political agenda. Different numbers arising from different definitions, different methods, and different organizational goals can be explained, but only if one has the time and inclination to listen to the explanation. Too often policymakers and elected officials have neither and so dismiss all tourism statistics as untrustworthy.

To summarize, the term and concept of 'tourism industry' is frequently used in an advocacy context to promote a diffuse constellation of thousands of uncoordinated small and medium-sized industries as part of something bigger, better organized, and worthy of greater political respect. Unfortunately, the strategy does not usually work well. Most members of the 'industry' neither view themselves as belonging to an industry nor act as if they do. Further, the proliferation of poor quality data and inconsistent definitions reinforces the impression by critics that claims concerning the importance of tourism cannot be trusted.

So where does this leave us? Is there any empirical or logical basis for referring to a 'tourism industry'? Is there any hope for consistency and credibility in tourism statistics?

Towards a new view of tourism

From a traditional macro-economic perspective, the answer to the question, 'Is tourism an industry?', must be 'no'. As noted previously, an industry is a group of businesses producing essentially the same type of commodity using essentially the same technology. The concept of a commodity involves a degree of judgement and the need to accept some level of generalization. For example, one can speak of a milk industry that produces milk and cream from cows (and, to a lesser degree, other mammals such as goats). This can be generalized into a dairy industry that also produces cheese, butter, yoghurt, and related products based on the production of milk and cream and the use of certain common processing techniques to produce these dairy products. This can be further generalized into an agricultural industry, encompassing not only dairy products but a wide range of food and fibre commodities produced from natural sources. Regardless of the level of aggregation, an industry is still characterized by the production of an identifiable generic product with a shared technology.

From this perspective, tourism is not an industry because the commodities produced by tourism businesses are viewed as heterogenous and produced via fundamentally different technologies. Passenger transportation on a commercial air carrier might be grouped with passenger transportation on a train under a transportation industry because both move people using mechanized vehicles, while hotels and private campgrounds might be grouped together into an accommodation industry. But there is no apparent commonality between moving people from place to place (transportation) and helping them stay still (accommodation).

On the other hand, transportation, accommodation, and other commodities are functionally linked in that they facilitate activities by people temporarily away from their home environment. The persistence of this functional linkage gives rise to the belief in a tourism industry. This 'no, but . . . ' response to whether tourism is an industry has led to a radically new way of measuring things like tourism that are not industries, but seem like they should be a satellite account. The term 'satellite' refers to the fact that this method involves the setting up of an accounting system that is a satellite or sub-set of a nation's system of national accounts. Before we discuss the notion of a tourism satellite account in greater detail, we need to establish a number of definitions.

Key definitions

The most fundamental definition in developing a tourism satellite account is that of 'tourism' itself. The tourism satellite account (TSA) is based on the definition of tourism adopted by the World Tourism Organization (1994a): tourism is the set of activities of a person travelling to and staying in places outside his/her usual environment for less than one year and whose primary purpose of travel is, other than the exercise, an activity remunerated from within the place visited.

The WTO definition thus considers tourism to be much more than just pleasure travel; certain forms of business travel, visiting friends and relatives (VFR), travel for health, and travel for religious purposes all constitute tourism. On the other hand, the definition excludes routine commuting to work, travel for temporary employment, travel by military and diplomatic personnel on assignment as well as their families, travel by refugees and nomads, cross-border travel by border workers during the course of their work, and travel by people changing their residence.

The definition of 'outside his/her usual environment' is, not unexpectedly,

from a drug store, or a family might travel by air (a tourism commodity) during their move to a new home (a non-tourism form of travel). From an industry perspective, tourism industries also supply non-tourism commodities. Many hotels sell snack foods, toiletries, and clothing. These are not tourism commodities because the demand for these products comes overwhelmingly from non-tourism sources. Conversely, some non-tourism industries supply tourism commodities, such as department stores that offer food services or car rentals.

A quantitative measure that helps to bring clarity to these patterns is the 'tourism ratio'. The tourism ratio is the percentage of total receipts in an industry that are attributable to tourism; in other words, it is the ratio between total tourism demand and supply. Every industry has a tourism ratio. For some, such as air transportation, this ratio will be over 90 per cent; for others, such as construction, it will be 0 per cent. How large does the tourism ratio have to be for an industry to be considered a 'tourism industry'? The answer is arbitrary. A useful rule-of-thumb is to consider tourism industries as any industry for which the tourism ratio is 15 per cent or above.

The complex mix of visitor purchases of tourism and non-tourism commodities from tourism and non-tourism industries is at the core of the conundrum of defining 'the tourism industry' and heightens the need for a new method such as a tourism satellite account to make sense of the complicated patterns of tourism supply and demand.

Structure of a tourism satellite account (TSA)

We can now turn to a fuller description of a tourism satellite account. As noted earlier, a TSA is based on a nation's system of national accounts. In a highly simplistic way, the system of national accounts might be thought of as a set of input/output (I/O) matrices in which every commodity is listed down rows and every industry across columns. Each cell in one matrix provides the value of every commodity produced in a given year by each industry. Another matrix presents the value of every commodity consumed by each industry. A TSA is a subset of these matrices.

Because tourism commodities are consumed by both visitors and non-visitors, and because visitors also consume non-tourism commodities, one cannot simply identify some set of industries and aggregate their statistics to describe a nation's tourism activity. Using data derived from consumer and business surveys, a TSA is able to identify the portion of outputs from both tourism and non-tourism industries consumed by visitors. The surveys must, of course, produce data that are accurate, reliable, and free from sampling bias. In practice, some of these surveys will be the same ones used to develop a nation's I/O tables; others will address individual travel patterns and consumer behaviour. Typically, such surveys are conducted by national statistical offices.

A TSA thus may be defined as an information system that combines, orders, and interrelates statistics describing all significant and measurable aspects of tourism within a framework that organizes tourism data according to the real world relationships from which they originate. A wide range of data can be contained in a TSA: (a) in the form of current and constant value, quantity, incidence rate, indices, and other statistics; (b) to describe various dimensions of tourism activity including economic, financial, managerial, social, demographic, and other information; and (c) relating to all facets of tourism such as consumers, producers, inputs, outputs, activities of organizations (such as government), factors affecting supply, and factors affecting demand.

Three principles underlay the creation of a TSA (Wells 1991):

1   The account must be based on recognized national economic accounting principles so that tourism can be meaningfully compared to other industries.
2   The account must include as wide a range of data as possible. The data should go beyond that normally associated with national accounting systems. For example, the satellite account can include socio-economic data, tourism volumes and visitor characteristics, and physical attributes and inventories. This extensive data base is needed to support diverse applications including policy analysis, industrial development, and market planning.
3   The account must provide user-friendly access to its database through a relational system that reflects the needs of diverse users. These needs include:

   A   Integrated statistics on a wide range of demographic, social, and economic factors reflecting the impact of tourism on the economy.
   B   Comparisons between tourism and other industries.
   C   Better representation of the organization of tourism, including explicit description of interrelationships among various tourism activities and the contribution of each industry (e.g. accommodation, transportation) to tourism.
   D   Information to support policies and other initiatives designed to ensure the nation maintains an appropriate share of international markets and supports a healthy domestic market.
   E   Guidance and discipline in the collection of tourism statistics including basic definitions, concepts, and classifications; identification of significant gaps in data; and identification and reconciliation of diverse data sources from both supply and demand data sources.
   F   Development of a standard of quality for collecting and reporting tourism statistics.

A full TSA consists of four separate layers of data, each representing a spreadsheet of information on tourism activities:

1   Money flows

    A   Tourism activity supply-side flows

         i   value of tourism commodities actually produced
        ii   value of governmental tourism-support programs
      iii   production costs of tourism commodities and support programs
      iv   value of imported tourism commodities

    B   Tourism activity demand-side flows

         i   visitors' (residents and foreigners) expenditures
        ii   tourism expenditures as intermediate consumption by industries and governments
      iii   expenditures on tourism-support programs as final consumption by governments and non-governmental organizations
      iv   investments by tourism and tourism-support governmental agencies

2   Quantity data

    A   Supply-side data

         i   quantities of tourism commodities (e.g. number of airline seats available)
        ii   quantity of inputs (e.g. number of employees)

    B   Demand-side data

         i   quantities of visitors (e.g. person-trips)
        ii   person-nights by type of accommodation (e.g. person-nights generated by hotels)

3   Characterization data

    A   Supply-side data

         i   description of outputs (e.g. price ranges of products in various industries; geographic locations of establishments; accommodation gradings)
        ii   description of inputs (e.g. age, sex, education of employees)

    B   Demand-side data

         i   description of visitors (e.g. age, sex, income, education, purpose of travel)
        ii   description of commodities purchased by businesses and the value of those commodities, reported by size, location, etc., of business

    iii description of commodities purchased by government and the value of those commodities, reported by level of government, spending unit, etc., of agency

4   Tourism planning and analysis information

    A   Supply-side data

        i quantity data

          a   stock of capital (e.g. number of hotels)
          b   stock of labour (e.g. size of labour force)

        ii financial data

          a   source of investment capital in tourism
          b   amount and structure of debt

        iii other

          a   unemployment rate
          b   GDP

    B   Demand-side data

        i quantity data

          a   potential visitors (e.g. total population, households by incomes) and ages of members

        ii financial data

          a   level of personal discretionary incomes, savings
          b   price indices

Level 1 data, financial flow information, are the foundation of a TSA. The conventions and discipline of national accounting are applied to this level's data before the TSA is expanded into other levels. This means, among other things, that total demand for a commodity (tourism plus non-tourism) must equal total supply; in other words, the demand and supply sides of the account must balance. Once this balance has been achieved, data at all other levels are tested to ensure they are consistent with the underlying financial flows. Discrepancies are examined and resolved; in practice, this usually means data at the higher levels will be adjusted as necessary to reflect the ratios and patterns contained in the financial flow accounts.

An empirical example

All of this may become a bit clearer with an empirical example. The volume of data at even level 1 of a TSA is too voluminous to report here, but the following summarized tables will illustrate the basic structure. To start, the example of the Canadian TSA illustrates the type of data sources required for a TSA. For the supply side of the account, revenues for each commodity category are required, as well as value-added by component for each industry. This information is obtained from worksheets used in developing the nation's I/O tables, data from special tabulations of business surveys, and other reference publications from Statistics Canada, such as reports on the consumer price index. Ultimately, this data is obtained from individual businesses under the Statistics Act, which means they have a legal obligation to respond to surveys and censuses. Demand-side data were drawn from five Statistics Canada surveys that collect information on travel by Canadian residents and international visitors. Micro-data files were used to extract or estimate the necessary figures for the TSA. One of the major sources of demand data is the Canadian Travel Survey (CTS), a quarterly survey that provides information on number of trips, purpose of travel, duration, distance, mode of transportation, spending, and other variables. The annual sample size is about 49,000. The CTS is administered as part of the Labour Force Survey, an omnibus survey that also falls under the Statistics Act. Data were also obtained from Statistics Canada's International Travel Surveys of Canadians travelling outside Canada and of international visitors entering the country. Finally, information was derived from other Statistics Canada surveys such as the Family Expenditure Survey, surveys of travel agents and tour operators, and worksheets for calculation of the consumer price index. More information on these sources may be found in Lapierre and Hayes (1994).

Table 3.1 is a summary of data related to supply and demand for various tourism commodities: transport, accommodation, food and beverage, other tourism commodities, and non-tourism commodities also purchased by tourists. The total supply and demand of the economy's commodities is also shown, for purposes of comparison. The first data column, 'domestic demand', is the sum of expenditures by business and government employees as well as private spending. 'Exports' is purchases by foreign visitors while in Canada. The third column, 'total tourism demand', indicates the sum of domestic demand and exports. 'Total domestic supply' is the total output of the commodity by Canadian industries. 'Imports' are purchases of foreign-produced commodities by Canadian residents travelling outside Canada. 'Tourism ratio' is the percentage of the total supply of a commodity purchased by visitors, or in other words, total internal demand (demand by Canadians travelling in Canada plus demand by international visitors travelling in Canada) divided by total domestic supply. For example, business, government, and private purchases of passenger air services for tourism purposes was $4.968 billion. International visitors purchased another $1.077 billion, for a total tourism demand of $6.044 billion. The value of all air passenger service produced in Canada – total domestic supply – was $6.566 billion. This yields a tourism ratio of 92.1 per cent.

Table 3.1 Tourism expenditures by commodity: data from the Canadian TSA

Tourism commodities	Domestic demand[a]	Exports[b]	Total tourism demand[c]	Total domestic supply[d]	Imports[e]	Tourism ratio[f]
	millions of dollars					%
Transport	10,960	2,333	13,294	31,372	3,092	42.4
Passenger air	4,968	1,077	6,044	6,566	1,650	92.1
Passenger rail	116	84	200	239	96	83.8
Passenger water	225	62	288	308	200	93.5
Interurban bus	315	87	402	456	213	88.1
Urban transit	20	14	34	1138	11	3.0
Taxis	275	122	398	1381	52	28.8
Vehicle rental	368	236	604	728	366	83.0
Vehicle repairs and parts	1,597	95	1,692	8893	96	19.0
Vehicle fuel	3,054	538	3,592	11,103	397	32.4
Parking	22	17	39	561	12	7.0
Accommodation	2,824	1,051	3,875	4,131	2,592	93.8
Hotels	1,940	764	2,704	2,959	1,858	91.4
Motels	455	134	589	617	308	95.4
Camping	169	44	213	255	113	83.6
Private cottages	125	41	166	—	150	—
Outfitters	90	45	135	153	82	88.2
Other	46	23	69	147	81	46.6
Food and beverage	4,186	1,499	5,685	22,206	1,838	25.6
Meals from:						
Accommodation	682	246	928	1,792	298	51.8
Food and beverage services	2,738	988	3,726	14,365	1,195	25.9
Other	106	38	144	519	46	27.7
Alcohol from:						
Accommodation	222	77	299	1,841	101	16.2
Food and beverage services	365	126	491	3,341	166	14.7
Other	72	25	96	348	32	27.7
Other tourism commodities	1,919	637	2,556	—	932	—
Recreation/ entertainment	1,389	617	2,005	7,235	877	27.7
Travel agencies	453	6	459	469	6	97.8
Convention fees	78	14	92	—	49	—

Table 3.1 continued

Tourism commodities	Domestic demand[a]	Exports[b]	Total tourism demand[c]	Total domestic supply[d]	Imports[e]	Tourism ratio[f]
			millions of dollars			%
Non-tourism						
commodities	3,915	1,014	4,930	—	1,988	—
Groceries	784	298	1,082	38,542	416	2.8
Alcohol from stores	91	33	124	7,595	35	1.6
Other	1,621	683	2304	—	937	—
Pre-trip	1,420	—	1,420	—	599	—
Total tourism						
expenditures	23,805	6,535	30,340	—	10,441	—
Total Economy	835,612	155,546	991,158	—	161,028	—

Source: Lapierre and Hayes (1994) 'The tourism satellite account'
Notes
a Domestic demand is the sum of purchases made by business and government employees travelling in Canada for which they are reimbursed by their employer, plus private spending by Canadian residents travelling in Canada.
b Exports is purchases made by foreign residents while visiting Canada.
c Total tourism demand is the sum of domestic demand and exports.
d Total domestic supply is the output of the specified commodity by Canadian industries.
e Imports are purchases of foreign-produced goods and services by Canadian residents while travelling outside Canada.
f The tourism ratio is the ratio between total domestic demand and total domestic supply; it may also be thought of as the percentage of the output of a commodity used by visitors.

One can note that the transportation sector is the largest general sector in tourism. The total supply of food and beverage services is also quite large, but only about 25 per cent of this output actually is consumed by visitors. One can also note that purchase of non-tourism commodities, such as groceries and alcohol from stores, represents a substantial volume of demand – $4.930 billion.

Tourism ratios vary from 1.6 per cent (alcohol from retail stores) to 95.4 per cent (motels). Only the accommodation sector has a tourism ratio over 50 per cent. Several transportation industries' ratios are over 80 per cent but the overall sectoral ratio is pulled down by very low ratios for urban transit, vehicle repairs and parts, fuel, and parking.

Table 3.2 presents industry-specific data, including labour income, GDP contributions, and employment and compensation data for a series of tourism industries as well as some non-tourism industries (for comparison purposes). Data for tourism industries has been weighted by the tourism ratio so that the estimates shown represent only that portion of labour income, GDP

Table 3.2 GDP and employment for tourism and non-tourism industries: data
from the Canadian TSA

Industry	Total labour income[b]	GDP at factor cost[c]	Persons employed[d]	Compensation per person employed[e]	GDP per person employed[f]
	(millions of dollars)		*(thousands)*	*(dollars)*	
1 Tourism industries[a]					
Transportation	2,600	4,141	77.3	34,600	53,600
Air	1,723	2,188	34.6	49,800	63,200
Rail	291	748	6.3	46,300	119,000
Water	118	142	2.5	47,200	56,800
Bus	222	289	10.3	21,800	28,000
Taxis	124	226	14.0	12,900	16,100
Vehicle renting	121	549	9.5	13,800	57,800
Accommodation	1,846	2,719	129.0	15,700	21,100
Hotels	1,407	1,957	92.7	15,700	21,100
Motels	252	474	21.3	15,700	21,100
Campgrounds	110	145	8.1	15,700	21,100
Other	77	141	6.9	15,700	21,100
Food and beverage	1,545	2,026	123.6	13,400	16,400
Licensed restaurants	817	1,050	65.0	13,400	16,200
Unlicensed restaurants	486	644	39.0	13,400	16,500
Take-out	208	289	16.9	13,400	17,100
Bars, nightclubs	34	43	2.7	13,400	16,000
Recreation and entertainment	419	943	25.2	20,100	37,400
Travel agencies	177	212	6.5	27,700	32,600
Total tourism industry	6,587	10,039	361.6	19,500	27,800
Non-tourism industries serving visitors	2,308	3,338	105.5	23,000	31,600
Total tourism-supported	8,895	13,377	467.1	20,300	28,600
2 Non-tourism industries	233,875	430,497	8,704.3	31,100	49,500
Agriculture	2,174	11,575	469.9	14,000	24,600
Fishing and trapping	339	1,201	41.5	22,400	28,900
Forestry	2,233	3,593	57.0	43,500	63,000
Mining and oil wells	7,130	20,499	144.9	49,900	141,500
Manufacturing	65,792	104,857	1,919.4	34,400	54,600
Construction	26,512	36,324	764.8	39,300	47,500
Transportation and storage	12,080	18,519	394.7	32,600	46,900

Table 3.2 GDP continued

Industry	Total labour income [b]	GDP at factor cost [c]	Persons employed [d]	Compensation per person employed [e]	GDP per person employed [f]
	(millions of dollars)		*(thousands)*	*(dollars)*	
Communications	8,022	14,611	207.4	38,900	70,500
Other utilities	4,106	17,207	106.1	38,800	162,200
Wholesale	19,690	27,795	596.3	33,800	46,600
Retail	23,706	32,440	1,421.5	18,100	22,800
Finance, insurance, real estate	25,585	82,805	662.4	56,900	125,000
Business/personal services	36,505	59,070	1,918.4	25,300	30,800
Total business sector	**242,770**	**443,874**	**9,171.5**	**30,500**	**48,400**

Source: Lapierre and Hayes (1994), 'The tourism satellite account'
Notes

a Data for tourism industries has been weighted by the tourism ratio so that the estimates shown represent only that portion of labour income, GDP contributions, and employment directly attributable to tourism.
b Total labour income includes wages, salaries, and benefits.
c GDP at factor cost includes wages, salaries, benefits, income of unincorporated business, and corporation profits before taxes, interest, dividends, and depreciation.
d Persons employed are 'full-time equivalents', excluding time-off for vacations or illness.
e Compensation per person employed is the sum of total labour income and the net income of unincorporated businesses (not shown) divided by the number of persons employed.
f GDP per person employed is GDP at factor cost divided by the number of persons employed.

contributions, and employment directly attributable to tourism. For example, the air transportation industry supports a tourism payroll of $1.7 billion and contributes $2.2 billion to the GDP (total labour income for the entire sector is about $1.8 billion and GDP at factor cost is about $2.4 billion). This industry employs 34,600 'persons' as a result of tourism (total 'persons' employed is closer to 38,000). It should be remembered that these 'persons' are full-time equivalents, not actual human beings. The total number of human beings employed in the airline sector during any given year will be larger. The difference is due to a number of employees who work only part-time or for only a portion of the year.

One also needs to recall the distinction between tourism industries (Table 3.2) and tourism commodities (Table 3.1). In the case of passenger airline service, virtually all air transportation (a commodity) in Canada is provided by airline companies (an industry). However, food and beverage services, for example, are provided not only by restaurants but by other industries such as hotels. A direct matching between commodities in Table 3.1 and industries in Table 3.2 is not possible.

Limitations and challenges of TSAs

Perhaps the greatest challenge associated with TSAs is their cost. The requirements for detailed data on supply and demand, tourism ratios, and all the other information a TSA requires is substantial. While most nations already have statistical systems in place that collect, process, and report data for aspects of their economic structure, these systems do not necessarily provide information at the necessary level of precision, on all the requisite variables, or on time cycles useful for the creation of a TSA. Further, national priorities, even for tourism statistics, do not necessarily support the creation of a TSA. For example, the ongoing development of the European Community and its needs for new national economic data, including tourism statistics, can result in deferring the development of a TSA indefinitely. In other nations, such as the United States, a lack of political will and the resulting under-funding of basic tourism data collection also acts as a barrier to developing a TSA. Concerns over cost will invariably be a reason for a nation not developing a TSA at this time. However, the response to the question, 'How much is this TSA going to cost us?', should be, 'How much will ignorance and mis-information about the true size of tourism cost us?'

One of the strengths of a TSA, the fact that it is based on a nation's I/O matrices, is also one of the TSA's limitations. National I/O matrices are updated infrequently because of the extraordinary amounts of work and data required for updating. In Canada, these matrices are updated only once every four years, and the results are released about four years after the reference year. As a result, tourism ratios, contributions to GDP, and other TSA-based data are dated. In fact, the data can be up to eight years old – too old to provide detailed information for business decisions.

Certain statistics such as the supply and demand of tourism commodities can, in principle, be obtained and released more frequently to provide industry with more timely data. Canada's national tourism indicators, for example, provide quarterly information on over 100 data series related to supply, demand, and employment that is no older than seventy-five days after the reference quarter. There are solutions to the lack of timeliness associated with TSA data, but the solutions require a long-term political and economic commitment for implementation.

TSAs normally offer only a national perspective. While this can be useful for purposes of national industry advocacy and economic planning, tourism fundamentally is a local industry in most countries. The cost and data requirements of TSAs mean that they will normally not be developed for sub-national levels, especially at the level of an individual municipality.

The structure of a TSA is shaped by a nation's SIC system. This implies that data are collected and reported only for industries as defined by a nation's existing SIC system. The classifications in these systems do not always provide the desired level of precision for tourism decision-makers and analysts. For example, public sector attractions such as national parks, museums, and art galleries are lumped into a general governmental category and excluded from the TSA because the public sector is not classified as an 'industry' in the context of the SIC.

Finally, the TSA involves definitions and ways of conceptualizing tourism that are different from those traditionally adopted by policymakers and decision-makers. For example, in Canada, some hospitality associations count everyone employed in the accommodation and the food services sectors as something called tourism-related employment. The TSA, on the other hand, discounts these estimates through the tourism ratio to reflect only the portion of industry revenues legitimately attributable to tourism expenditures. The differences in the two perspectives results in substantially different estimates of employment and, as a result, lingering mutual doubts about the credibility of each other's data.

In sum, the TSA is a powerful new tool for describing and measuring the structure and magnitude of the tourism 'industry' but it requires a substantial investment in technical expertise, data collection, and money to produce and maintain. It also requires a period of education of industry leaders and others so they can interpret the results of a TSA.

Closing observations

The TSA is not a definition of tourism; it is a tool for measuring and describing economic activity directly attributable to tourism. The definition of tourism must be established prior to developing a TSA. That definition, for the sake of consistency and comparability, should be WTO's definition: the set of activities of a person travelling to and staying in places outside his/her usual environment for less than one year and whose primary purpose of travel is, other than the exercise, an activity remunerated from within the place visited. Note, however, this is a demand-side definition; that is, it defines tourism in terms of characteristics of consumers.

Thus, a TSA implicitly defines the tourism 'industry' as a demand-side concept, contrary to the convention of defining industries from a supply-side perspective. This is a dangerous conceptual situation. It is as if, for example, the motor vehicle industry were defined in terms of what a driver is, or if health care services were defined in terms of who a sick person is. The risk is that one could begin to define all sorts of meaningless industries solely in terms of groups of consumers. One might, as Leiper warned, absurdly define a 'red-heads industry' (Leiper 1990a: 602) consisting of those firms producing commodities consumed by red-headed individuals. Because commodities consumed by red-headed individuals are also consumed by brunettes, blondes, greys, and bald-headed individuals, we could also develop a 'red-head ratio' to reflect the portion of commodity sales attributable to the red-headed industry.

This nonsensical situation can be avoided if we change our conception of tourism products from a bundle of heterogenous commodities consumed individually by visitors in the course of a trip to a more generic and singular tourism 'product'. The commodities that red-heads consume do not share any feature or purpose in common other than the colour of the hair of the consumer. Tourism commodities, on the other hand, are linked in that they all support the production of a tourism experience. Such a conceptual shift helps us move towards re-establishing the 'tourism industry' as a valid supply-side concept.

The question is now, 'Is there a generic tourism product?' Before answering that question, it may be helpful to consider what constitutes a product. Kotler (1984: 463) offers a useful definition: a product is 'anything that can be offered to a market for attention, acquisition, use, or consumption that might satisfy a want or need. It includes physical objects, services, persons, places, organizations, and ideas'. This definition is relevant because it explicitly argues that products are not just physical objects or services, but also the other 'material' that constitutes tourism: persons, places, organizations, and ideas.

With respect to tourism, a variety of authors (Medlik and Middleton 1973; Wahab et al. 1976; Schmoll 1977; Middleton 1989; Gunn 1994; Jefferson and Lickorish 1988; Lewis and Chambers 1989) long have argued that the tourism product can be conceptualized as an experience involving all aspects of a trip, including physical objects, services, and the 'other stuff' referred to in the previous paragraph. However, none of these authors has adequately specified all the components of a generic tourism product as well as a general production process (technology) that produces the tourism product.

S. Smith (1994) proposes a hypothetical structure for a tourism product as well as a conceptual tourism production process. In brief, Smith's model of the generic tourism product (Figure 3.1) identifies five components that are present, in varying degrees, in every tourism product (or experience): (a) some physical element such as a resort hotel; (b) service; (c) hospitality (which refers to attitude, whereas service refers to the technical performance of some skill); (d) the individual having some sense of choice or selection (even if it is only the decision to travel); and (e) a sense of involvement or engagement in the act of travel. The relative importance of each of these elements will vary depending on the specific tourism experience but, Smith argues, all must be present for a visitor to have a satisfactory tourism experience. The relevance of this model to the current discussion of the nature of the tourism industry may be a bit clearer when we consider the other half of Smith's conception of the tourism product: the tourism production process.

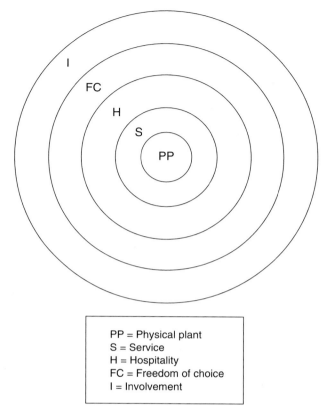

Figure 3.1 The generic tourism product

The creation of a tourism product involves four steps (Table 3.3). The process begins with primary inputs or resources such as land, labour, and capital. These are transformed into intermediate inputs (physical facilities) such as parks, resorts, and convention centres. Further processing (additional inputs of labour and management) results in the creation of intermediate outputs: services such as artistic performances, guided tours, festivals and events, and accommodation and food services. In the final stage of the production process, the visitor takes the intermediate outputs and processes them into final outputs or experiences.

It is this last stage of the production process that is a distinguishing feature of the tourism industry: the consumer is an integral part of the production of the tourism product. In fact, the consumer actually produces the final tourism product – the experience. Unless the consumer actually uses the intermediate outputs in the course of being a visitor, there is no experience, no final tourism product. Without the consumer, the production process ceases at the level of

Table 3.3 The tourism production process

Primary inputs → (resources)	Intermediate inputs → (facilities)	Intermediate outputs (services) →	Final outputs (experiences)
Land	Parks	Park interpretation	Recreation
Labour	Resorts	Guide services	Social contacts
Water	Transportation modes	Live performances	Education
Agricultural produce	Museums	Souvenirs	Relaxation
Fuel	Craft shops	Conventions	Business contacts
Building materials	Convention centres	Accommodation	Memories
Building	Hotels	Meals and drinks	Festivals and events
Capital	Restaurants		
	Rental car fleets		

Source: S. Smith (1994)

intermediate outputs. A TSA describes the value of and relationships among these intermediate outputs. The TSA stops at this stage because existing methods and concepts are not designed to measure any aspect of the subjective experience of tourism – the final product. However, even though the current statistical system stops short of quantitatively describing the final tourism product, one can still recognize in principle the existence of a generic product, a tourism experience. Note that this logic could not be meaningfully applied to situations such as Leiper's deliberately absurd notion of a red-heads industry. Tourism may legitimately be considered as an experience composed of a combination of individual commodities; the same cannot be said of being a red-head.

Tourism clearly is not an industry in the conventional sense of the word. The production process used to create a tourism product is not a conventional technology. The tools used to describe and measure the size of tourism as an area of economic activity are not conventional tools. Nonetheless, tourism can be considered as an industry: perhaps a 'synthetic industry', a 'matrix industry', or a 'composite industry' – but still an industry.

4

THE TOURISM PRODUCTION SYSTEM

The logic of industrial classification

Wesley Roehl

Introduction

One of the challenges to the study of tourism is its definition. Unlike many other economic activities, there is no single definition of tourism that is universally accepted. One approach, discussed by Smith in Chapter 3, is to approach tourism through a system of satellite accounts. A second approach, and the one featured in this chapter, is to study tourism by understanding how it fits within the context of existing systems for classifying economic activity.

Specifically, this chapter is organized in five sections. The first will consider the challenges raised to the study of tourism geography by the economic structure of tourism. The second section will discuss how tourism-related activities fit within the logic of industrial classification systems. In the third section, a major revision of industrial classification affecting the United States, Canada, and Mexico will be discussed and tourism-related industrial classifications in this new North American industry classification system (NAICS) will be identified. This will be followed by a section that illustrates the use of tourism-related industrial classification codes in a case study of the tourism product life cycle and economic restructuring over time. In the final section of this chapter, areas for further research will be identified.

Tourism geography and the economic geography of tourism

As noted by S. Britton (1991) and Ioannides (1995), the study of tourism has been handicapped by inattention to the supply side. These authors argue that understanding tourism requires an understanding of the tourism production system. The tourism production system represents the mix of businesses and other organizations that provide tourism services. This mix need not be static, it can change over time. Furthermore, changes to the tourism production system are likely to be affected by restructuring in the broader economy. S. Britton (1991: 475) crystallized this perspective when he wrote 'we need a theorization that explicitly recognizes, and unveils, tourism as a predominantly capitalistically organized activity driven by the inherent and defining social dynamics of that system, with its attendant production, social, and ideological relations'.

This call to understand better the tourism production system fits into the broader recognition that services, a vital part of the modern economy, are not as well conceptualized and measured as they should be. In part this may be due to academic and organizational inertia. But the definition and measurement problems that accompany services also contribute to this problem. The conceptual organization of the economy, classifying activities into primary or extractive activities, secondary or transformative activities, and tertiary or service activities (Clark 1940; Fisher 1935) dates from a time when manufacturing was seen as the engine driving modern economies. Services represented a residual category; if an activity did not produce a raw material, or transform that raw material into a finished product, it was assigned to the residual service category. Despite the fact that services had become the dominant sector in the United States and Canadian labor force by 1960, efforts to understand the nature of services, develop detailed definitions of services, and measure services comprehensively have lagged behind research in other sectors of the economy (Kellerman 1985).

Given the relative lack of prestige and attention directed to the service sector of the economy, it is not surprising that tourism has been ill defined. Leiper (1993a) illustrates one aspect of this problem when he identifies three sets of meaning for the term tourist. Similar problems arise when examining the supply side of the tourism equation. Travelers, whether traveling for business, to visit friends and relatives, or for vacation/holiday typically purchase services from a number of different types of firm. Some of these firms may generate all or most of their revenues from travelers. Others may serve a mix of travelers and local residents, while still others may get most of their revenue from residents, only occasionally serving a tourist.

Because of these characteristics, a *tourism industry* would inherently represent a market- based grouping of economic activities. In a market-based grouping of economic activities, activities are categorized by both their output or product and by the characteristics of their market. This requires disaggregating measures of output such as revenue or receipts by type of customer (Economic Classification Policy Committee 1993 and 1994b). Sales to a tourist would be considered part of the tourism industry, while sales to a resident would not be included. Inherent to this approach is the idea of separating revenue based on whether it comes from within the local geographic area or from outside the local geographic area. However, this leads to other problems. Take the activities of a travel agency. Typically, local residents spend locally-earned money at a local travel agency. But using a market-oriented approach to defining tourism, should these expenditures be counted as revenue for the tourism industry in that local area? One way to implement a market-oriented definition of tourism implies that tourism revenues are being generated by individuals coming into a specific area from outside. This would suggest that revenue received in a travel agency from local residents should not be included as part of tourism! Additionally, a market-oriented approach, while serving many uses, also makes serious demands on data collection agencies and on the establishments providing the data in that they would have to disaggregate their measures of output by type of customer (Economic Classification Policy Committee 1994b).

In contrast, a production-oriented approach to industry definition identifies industries based upon their outputs and production functions (Economic Classification Policy Committee 1994a and 1994b). While this approach often contributes to difficulties in identifying services, it is consistent with definitions for other sectors of the economy and it fits into the current data collection regime.

At present, there seems to be an emerging consensus that tourism is not an industry, if industry is to be used in a manner consistent with how it is used in other fields. Market-oriented definitions of the tourism industry are being provided through the development of tourism satellite accounts. As mentioned in this chapter's introduction, Smith's contribution in Chapter 3 of this book provides an excellent summary of this matter (see also Meis *et al.* 1996). The remainder of this chapter will investigate how tourism fits into the existing industry definition and data collection framework. In the next section the process of industrial classification will be discussed and the place of tourism within the classification system analyzed.

Tourism and the logic of industrial classification

Governments, businesses, and international organizations have long recognized the benefits of classifying establishments based upon their economic activity. These classification systems aid understanding: (a) the structure of an economy at a point in time, (b) the effect on specific parts of the economy of domestic policy changes, (c) the effect of international events and competition on the economy, etc. There are a number of different coding schemes, for example the standard international trade classification of the United Nations, the harmonized system from the Customs Cooperation Council, and the standard industrial classification scheme of the European Union. However, most of these approaches share similarities in structure and face many of the same problems when it comes to operationalization.

The basic goal in any of these systems of classifying economic activity is to create a logical structure that is both parsimonious and descriptive. In other words, the goal is to create a typology organizing/describing activities such that activities in a common category are more similar to one another than they are to activities in other categories. At one extreme, every economic activity could be assigned its own category. This would achieve homogeneity within categories and heterogeneity among categories but at the price of a very large number of categories. This would violate parsimony. At the other extreme, all activity could be placed in a single category. This would be a parsimonious, but not very descriptive, solution. Typically, compromise is achieved through the use of a hierarchical structure. This approach achieves both parsimony, in that a small number of categories can describe an entire economy, and description, in that detail can emerge through the use of a detailed hierarchical structure within each of a relatively small number of major categories.

A good example of these typologies is provided by the US standard industrial classification (SIC) system and the new North American industry classification system (NAICS) which was implemented by the United States, Canada, and Mexico in 1997.

The SIC is a system designed for statistical purposes and used by agencies of the US Government that collect or publish data by industry. It is based on the following four principles:

1 The classification is organized to reflect the structure of the US economy. It does not follow any single principle, such as end use, nature of raw materials, product, or market structure.

2 The unit classified is the establishment. An establishment is an economic unit that produces goods or services – for example a farm, mine, factory, or store. In most instances, the establishment is at a single physical location and is engaged in one, or predominantly one, type of economic activity. An establishment is not necessarily identical with a company or enterprise.

3 Each establishment is classified according to its primary activity. Primary activity is determined by identifying the predominant product, or group of products, produced or handled, or service rendered.

4 An industry (four-digit SIC) consists of a group of establishments primarily engaged in the same activity. To be recognized as an industry, such a group of establishments must meet certain criteria of economic significance, as described in Section D (Executive Office of the President 1987: 699).

This system divides the economy into divisions (such as mining, retail trade, and public administration) representing broad sections of the economy (Table 4.1). Each division is further divided into major groups. Major groups (two-digit level of detail) are further divided into industry groups (three-digit level of detail). Industry groups represent aggregations of highly similar industries. In the SIC system the industry (four-digit level of detail) is the base unit of economic classification. For example, within the division services is major group 70, *hotels, rooming houses, camps, and other lodging places*. In this major group, *hotels and motels* are designated as an industry group (SIC 701). Within this industry group is a single industry, *hotels and motels*, SIC 7011 (Table 4.2). Establishments found in SIC 7011 include auto courts, bed and breakfast inns, cabins and cottages, casino hotels, hostels, hotels, except residential, inns furnishing food and lodging, motels, recreational hotels, resort hotels, seasonal hotels, ski lodges and resorts, tourist cabins, and tourist courts.

Table 4.2 illustrates the hierarchical structure typical of most industrial classification systems. It also shows how this system can both contribute to understanding the tourism production system and can confuse the issue. Using Table 4.2 and the 1987 US SIC as an example, three of the five industry

Table 4.1 Structure of the 1987 revision of the US standard industrial classification

Division	Description
A	Agriculture, forestry, and fishing
B	Mining
C	Construction
D	Manufacturing
E	Transportation, communication, electric, gas, and sanitary services
F	Wholesale trade
G	Retail trade
H	Finance, insurance, and real estate
I	Services
J	Public administration
K	Non-classifiable establishments

Table 4.2 US SIC major group 70: hotels, rooming houses, camps, and other lodging places

Division	Major group (two digit)	Industry group (three digit)	Industry (four digit)	Definition
I	70	701	7011	Hotels and motels
I	70	702	7021	Rooming and boarding houses
I	70	703	7032	Sporting and recreational camps
			7033	Recreational vehicle parks and campsites
I	70	704	7041	Hotels and lodging houses, on membership basis

groups fit into the tourism production system. However, one of the industry groups (SIC 704), containing industry 7041 *organization hotels and lodging houses, on membership basis*, may not be part of the tourism production system. Similarly, whether the industry group that includes rooming and boarding houses (SIC 702) is part of the tourism production system would depend on the nature of the patrons at these establishments. For a given geographic area, if information is available for each four-digit industry or three-digit industry group in major group 70, then the existence of SIC 7041 and SIC 7021 is not a problem. However, if the given geographic area has few firms in major group 70 or has few firms in some of the industry groups or industries making up major group 70, data may only be published for the major group as a whole, and not for the more disaggregated categories. This raises problems for tourism researchers. If they accept major group 70 as part of the tourism production system, the potential exists for mismeasurement. It will be up to the researchers to determine the magnitude of this mismeasurement and the implications it has for their research.

Ideally, establishments are classified based upon their primary activity or product. However, since the SIC codes have periodically been revised and, since one of the defining principles of the system was to reflect the structure of the US economy, a number of different criteria have been used to classify economic activities. These include the typical product or activity produced, but also the end use of the product, the nature of the raw material(s) used in the production process, and the market structure. Thus in the 1987 revision to the SIC, for example, six separate industries produced chairs. An establishment producing chairs was classified into one of these industries based upon whether the chair was wood or metal, upholstered or not, and intended for home or office use (Norwood and Klein 1989).

Inertia has also played an important role in the evolution of the SIC. When the United States began organizing the SIC, manufacturing was the dominant economic activity. It should be no surprise then, that even in the 1987 revision, much more detail was available for manufacturing than was available for services. In part, this is also due to the criteria for changing classifications. First, changes that fit easily into the existing system were more likely to occur than were changes to the basic two-digit and three-digit structure of the SIC. Second, historical continuity was a major goal. Third, to be recognized as an industry, a group of establishments had to achieve an economic significance to the economy of at least 20 per cent of the average of industries in that division. This economic significance was evaluated using a weighted index of number of establishments, number of employees, payroll, value added, and value of shipments. Since many services were located in a very broadly defined, highly heterogeneous division it was often more difficult for a new service to achieve this 20 per cent criteria than it was for a new manufacturing industry. Fourth, to be recognized as a new industry, that industry's output must consist mainly of the goods and services defining that industry and account for the bulk of that type of good or service produced by all industries. Finally, other criteria considered in identifying new industries included efforts to accommodate rapidly growing industries and to bring the US system into closer compatibility with other major classification systems, such as that of the United Nations (Executive Office of the President 1987).

Within the SIC revision of 1987 there was no division, major group, industry group, or industry labeled tourism. However, some activities that are part of the tourism production system could be identified using the SIC. Four digit industries such as SIC 4722 *arrangement of passenger transportation*, SIC 7011 *hotels and motels*, SIC 7033 *recreational vehicle parks and campsites*, and SIC 7996 *amusement parks* would likely fit in most definitions of the tourism production system. Most, if not all, of the output generated by establishments in these four-digit Industries is consumed by tourists. Whether major group 84, *museum, art galleries, botanical, and zoological gardens*, is part of the tourism production system would depend on the area under study and the definitional perspective used by the researcher. But, on the other hand, a number of activities that are clearly part of the tourism production system could not be identified using the 1987 SIC. For example, sightseeing services are unarguably part of tourism. But information on sightseeing services is not recoverable in the 1987 SIC. Sightseeing airplane services is just one activity included in SIC 4522 *air transport, non-scheduled* along with activities that are emphatically not part of tourism such as air ambulance service and non-scheduled air cargo carriers. Similarly, sightseeing busses are located in SIC 4119 *local passenger services, not elsewhere classified* together with activities such as hearse rental, with driver.

Overall, the 1987 revision to the SIC system left tourism researchers with opportunities and challenges. Opportunities existed in that some of the core activities of the tourism production system such as lodging, tour operation, and travel agencies were uniquely identifiable in the system at the major group, industry group, or industry level. However, a number of challenges existed for the tourism-oriented user of this data. First, the internal logic behind classification was not consistent. Rather, it represented a variety of perspectives for classifying establishments. Second, history and inertia had resulted in many service-related classifications that included dissimilar economic activities. Classification had not kept pace with the evolution of the economy in general or the development of the tourism production system, in particular.

Interestingly, many of these concerns were identified by other users of the US industrial classification system. This dissatisfaction, coupled with the signing of the North American Free Trade Agreement, led to the establishment of a new set of codes that will allow the Mexican, Canadian, and US governments to share a revised and more detailed set of codes that more accurately reflect the structure of North American economies at the dawn of the twenty-first century. In the next section, the new North American industrial classification system (NAICS) will be discussed, some issues related to implementing NAICS identified, and NAICS codes identifying establishments that are part of the tourism production system illustrated.

NAFTA and SIC codes – the NAICS

Beginning in 1992 the US Office of Management and Budget established a committee to examine economic classification. This committee was charged with developing an improved industrial classification system to be in place by 1997 (Office of Management and Budget 1994 and 1996). This task was broadened as Statistics Canada and Mexico's Instituto Nacional de Estadistica, Geografia e Informatica (INEGI) entered into the process after NAFTA was signed. Following public input, consultation with data consumers and industry groups, and discussion with Statistics Canada and INEGI, a new industrial classification system was developed that would apply to all three countries.

There are four basic principles that serve as a foundation for this system. First, it is a production-oriented framework. In this framework, the production technology of an industry, such as the production process, material used, type of labor employed, or some combination of process, material, and labor, define the industry. To state this another way, the producing units within an industry share a production function. Establishments belonging to different industries will have different production functions. Therefore, NAICS industries reflect different production processes and production technologies. Furthermore, an industry is a grouping of economic activities. Although the products of economic activities are included in the industry definition, industries do not represent groupings of products.

A second goal of NAICS was to correct the historical bias toward manufacturing and the under-identification of service industries. Development of classifications for '(a) new and emerging industries, (b) service industries in general, and (c) industries engaged in the production of advanced technologies' was emphasized (Office of Management and Budget 1996).

A third goal was time series continuity and the adjustment among classifications to develop a common system for Canada, Mexico and the United States. It was recognized, however, that time series continuity would be easier to maintain in older, more traditional manufacturing and primary production sectors. Finally, a fourth goal was to develop a system compatible with the two digit level of the United Nations standard industrial classification of all economic activities.

These four principles represent both continuity with, and change from, the logic underlying the older SIC system. Continuity exists due to the continued emphasis on the establishment as the unit of measurement. Similarly, both approaches stress classification through identification of an establishment's primary activity. The systems differ, however, in that the NAICS explicitly adopted the production function as its organizing principle, abandoning the more eclectic approach featured in the SIC.

The structure of the new NAICS expands classification from the former US SIC codes' four-digit level of disaggregation to a six-digit level of disaggregation. The first two digits of the new system identify sectors of the economy (Table 4.3). By expanding this part of the classification from one to two digits, a more detailed and more realistic description of the economy is possible. The present NAICS consists of twenty-five sectors, compared to eleven divisions in the 1987 US SIC. While there was a single division representing services in the 1987 US SIC, the NAICS has seven sectors covering service industries. The third digit in the NAICS identifies the specific subsector, the fourth digit identifies the industry group, and the fifth digit designates the NAICS industry. Furthermore, while the basic building block of the system is the five-digit NAICS industry, each of the three nations is authorized to use a sixth digit in those cases where economic specialization within a country justify the added detail. Overall, the expansion from the 1987 SIC four-digit system to the NAICS six-digit system does not generate more classification detail. Instead it reflects more detail at the sector level of the economy and more detail at the individual country industry level.

Table 4.3 Structure of the North American industrial classification system

Two-digit sector	Description
11	Agriculture, forestry, fishing, and hunting
21	Mining
22	Utilities
23	Construction
31–33	Manufacturing
42	Wholesale trade
44–45	Retail trade
48–49	Transportation and warehousing
51	Information
52	Finance and insurance
53	Real estate and rental and leasing
54	Professional, scientific, and technical services
55	Management of companies and enterprises
56	Administrative and support, waste management, and remediation services
61	Educational services
62	Health care and social assistance
71	Arts, entertainment, and recreation
72	Accommodation and food services
81	Other services (except public administration)
92	Public administration
99	Unclassified establishments

A good example of this detail is found in the NAICS Industry *passenger car rental*. This activity is part of NAICS sector 53, *real estate and rental and leasing*. Under the industry group 5321, *automotive equipment rental and leasing*, are a variety of activities. *Passenger car rental and leasing* is coded as NAICS industry 53211. However, in the United States passenger car rental and passenger car leasing are both sizeable economic activities that differ in their production functions. Therefore, in the application of NAICS in the United States a pair of six-digit US national industries are identified, 532111, *passenger car rental*, and 532112, *passenger car leasing*. However, compatibility among the three countries is maintained since US national industries 532111 and 532112 can be aggregated back to the five-digit NAICS industry level, 53211, present in Mexico and Canada (NAICS Committee 1995e).

Among those who will benefit from NAICS are researchers interested in the geography of tourism. Among the new industries are several that would appear to be part of the tourism production system (Table 4.4). The next few paragraphs will identify and discuss briefly some of the NAICS codes of particular interest in understanding tourism.

Considerable detail has been added in the transportation industries. Scheduled passenger air transportation has been separated from scheduled freight air transportation. Inter-urban and rural bus lines have been separated from urban transit systems. Perhaps the most exciting change in transportation coding is the identification of *scenic and sightseeing transportation* as a sub-sector of the transportation sector. Detailed coding is available for land, water, and other scenic and sightseeing transportation.

The identification of *scenic and sightseeing transportation* as a NAICS industry results from applying the product function approach to industrial classification. Scenic and sightseeing transportation differs from other forms of transportation in both the nature of the product and in the production process. In contrast to other transportation activities, efficiency is not emphasized. Instead obsolete equipment such as steam trains and horse-drawn carriages are used to produce a product in which ambience is a major product feature (NAICS Committee 1995a).

As mentioned earlier, the NAICS sector *real estate and rental and leasing* contains the six-digit US national industry classification for passenger car rental. This has been separated from the activity of passenger car leasing.

Administration and support, waste management, and remediation services (NAICS sector 56) contains the four-digit industry group *travel arrangement and reservation services*. This category includes NAICS industries *travel agencies* (56151), *tour operators* (56152), and *other travel arrangement and reservation services*. As a detailed six-digit US national industry, *convention and visitor bureaus* are identified as a separate industry (NAICS Committee 1995c).

Depending on how tourism is defined, the NAICS sector *arts, entertainment, and recreation* (Sector 71) contains a number of activities that may or may not be appropriate for the tourism production system. Three sub-sectors (three

Table 4.4 NAICS codes that describe the tourism production system at the NAICS industry and NAICS US national industry level

---

48–49 Transportation and warehousing[a]

48111	Scheduled air transportation[b]
481111	Scheduled passenger air transportation
48521	Interurban and rural bus lines
48531	Taxi service
48711	Scenic and sightseeing transportation, land
48721	Scenic and sightseeing transportation, water
48799	Scenic and sightseeing transportation, other
48811	Airport operations
488111	Air traffic control
488112	Air operations (except air traffic control)
48819	Other support activities for air transportation

53 Real estate and rental and leasing

53211	Passenger car rental and leasing
532111	Passenger car rental

56 Administrative and support, waste management, and remediation services

56151	Travel agencies
56152	Tour operators
56159	Other travel arrangement and reservation services
561591	Convention and visitors bureaus
561599	All other travel arrangement and reservation services

71 Arts, entertainment, and recreation

71111	Theater companies and dinner theaters
71112	Dance companies
71113	Musical groups and artists
71119	Other performing arts companies
71121	Spectator sports
711211	Sports teams and clubs
711212	Race tracks
711219	Other spectator sports
71131	Promoters of performing arts, sports, and similar events with facilities
71132	Promoters of performing arts, sports, and similar events without facilities
71141	Agents and managers for artists, athletes, entertainers, and other public figures
71151	Independent artists, writers, and performers
71211	Museums
71212	Historical sites
71213	Zoos and botanical gardens
71219	Nature parks and other similar institutions
71311	Amusement and theme parks
71312	Amusement arcades

Table 4.4 continued

71321	Casinos (except casino hotels)
71329	Other gambling industries
71391	Golf courses and country clubs
71392	**Skiing facilities**
71393	Marinas

72 Accommodation and food services

72111	**Hotels (except casino hotels) and motels**
72112	**Casino hotels**
72119	**Other traveler accommodations**
721191	**Bed and breakfast inns**
721199	**All other traveler accommodations**
72121	**RV (Recreational Vehicle) parks and recreational camps**
721211	**RV parks and campgrounds**
721214	**Recreational and vacation camps**
72131	**Rooming and boarding houses**
72211	Full-service restaurants
72221	Limited-service eating places
722211	Limited-service restaurants
722212	Cafeterias
722213	Snack and non-alcoholic beverage bars
72231	Foodservice contractors
72232	Caterers
72233	Mobile caterers
72241	Drinking places (alcoholic beverages)

Notes:

a NAICS sectors are italicized.

b NAICS industries and NAICS US national industries that serve primarily travelers are indicated with bold type; other listed NAICS industries and NAICS US national industries may serve travelers but this relationship is likely to be specific to the geographic context.

digits) are present: *performing arts, spectator sports and related industries, museum, historical sites, and similar institutions, and amusement, gambling, and recreation industries*. As an example of the types of changes inherent in the NAICS, the 1987 SIC 7993 *coin-operated devices* has been split to form two new five-digit NAICS industries, *amusement arcades* and *other gambling industries* (NAICS Committee 1995d).

Recognition of the role played by gambling in the modern economy continues in NAICS sector 72, *accommodations and food services*. Casino hotels are defined as an NAICS industry at the five-digit level separate from other hotels and motels. Bed and breakfast inns have also been identified as a US national industry. Additional detail is also provided in the description of RV (Recreational Vehicle) parks and recreational camps. Similarly, additional detail has been added for the United States in the three-digit subsector *foodservices and drinking places* (NAICS Committee 1995b).

Table 4.4 should not be considered the only possible definition of the tourism production system that can be developed from the NAICS. Clearly judgement should play an important role in defining the tourism production system. If, for example, taxis in an area service primarily visitors to that area, then the tourism production system for the area should include taxis. However, researchers located in North America, or interested in understanding tourism in North America, should take advantage of the new NAICS and its added detail to define tourism as it applies to their particular study.

To reinforce the idea that the utility of using existing industrial classification systems to describe change in tourism is high, an example is given in the next section. This section will review briefly a pair of studies by Roehl and Fesenmaier (1988a and 1988b) that used the US SIC to define tourism. These authors were interested in change in the structure of tourism over time. Their work will illustrate how industrial classification systems can be used to understand the tourism production system, the evolution of tourism production systems over time, and the relationship of changes in the tourism production system to restructuring in the broader economy.

Using SIC codes to understand changes in the tourism production system

So far, this chapter has discussed some of the issues surrounding the definition and organization of economic classification systems. Specifically, the US SIC and the new NAICS coding schemes have been discussed. In this section, a pair of studies that used SIC codes to define and describe change in the tourism production system will be reviewed as an example of the potential application of this approach to the study of tourism.

In a pair of studies, Roehl and Fesenmaier (1988a and 1988b) examined change in the number and mix of tourism establishments in Texas. Using SIC codes (Executive Office of the President 1972), nine economic activities that serve primarily travelers and tourists were defined as the tourism production system (Table 4.5). These activities included four-digit SIC industries such as *hotels, tourist courts, and motels* (SIC 7011), three-digit industry groups (SIC 451 *air transportation* and SIC 458 *Fixed facilities and services related to air transportation*) summed together to form a single entry, and a two-digit SIC major group, *museums, art galleries, botanical, and zoological gardens* (SIC 84). Data from the US Department of Commerce's County Business Patterns for the years 1974, 1976, 1978, 1980, and 1982 were used to describe tourism in each of Texas's 254 counties. To avoid problems with data suppression and seasonality of employment data, the number of establishments was used to characterize the size and diversification of the tourism production system.

Both studies followed the strategy of describing the tourism production system by identifying the dominant tourism function in each county, distinctive tourism functions in each county, and measuring functional specialization

Table 4.5 SIC codes used to define the tourism production system

SIC code	Description
451	Air transportation, certified carriers
458	Fixed facilities and services related to air transportation
4722	Arrangement of passenger transportation
48	Museums, art galleries, botanical, and zoological gardens
7011	Hotels, tourist courts, and motels
7032	Sport and recreational camps
7033	Trailer parks and campsites for transients
7512	Passenger car rental and leasing, without drivers
7996	Amusement parks

Source: Roehl and Fesenmaier (1988a and 1988b)

in each county (Maxwell 1965; Mulligan and Reeves 1986). The dominant tourism function was the type of tourism activity with the most establishments in a given county. Distinctive tourism functions were those activities present in a county at levels more than two standard deviations above the mean for all Texas counties during the study period (e.g. if the mean number of establishments of type x per county was five, with a standard deviation of two, then those counties having more than nine establishments of type x were identified as having distinctive levels of type x establishments). Functional specialization describes the diversity of tourism establishments present in a county. This was measured using a standardized entropy coefficient: $H = -\Sigma p_i \ln p_i / \ln k$, where p_i is the percentage of the total tourism establishments accounted for by category I tourism establishments, ln is the natural log, and k is the number of categories of tourism establishments (Garrison and Parker 1973; Hackbart and Anderson 1975; Shannon and Weaver 1949). The standardized entropy coefficient will equal one only when all of the establishments are concentrated in a single category. It will equal zero where there are an equal number of establishments in each category.

In their first study (Roehl and Fesenmaier 1988a) 110 counties were identified in which there were no tourism establishments in 1974 (Table 4.6, Figures 4.1 and 4.2). However, by 1982, only eighty-six counties were without tourism establishments. During this period, hotels were the type of tourism activity that most frequently had the highest number of establishments in Texas counties. Hotels were the dominant activity in between seventy-two and eighty-seven Texas counties during this period. An interesting trend was the growth in the number of establishments providing the arrangement of passenger transportation. While this was the dominant activity in twenty-four counties in 1974, by 1982 it was the dominant activity in sixty counties.

Table 4.6 Number of Texas counties in which one of the eight tourism activities accounted for the largest number of establishments, 1974–82

Economic activity	Year					
	1974	1976	1978	1980	1982	Total
Air Transportation (451+458)	1	2	0	0	1	4
Arrangement of passenger transportation (4722)	24	21	28	41	60	174
Hotels (7011)	72	79	87	80	76	394
Sport and recreational camps (7032)	17	19	15	19	22	92
Trailer parks and campsites (7033)	14	13	9	7	4	47
Passenger car rental (7512)	11	14	13	7	4	49
Amusement parks (7996)	5	3	2	2	1	13
Museums, etc. (84)	0	0	0	0	0	0
None[a]	110	103	100	98	86	497
Total	254	254	254	254	254	1,270

Note:
a These counties did not contain any travel and tourism establishments.

In contrast, the number of counties with distinctive levels of tourism activities was consistent throughout this time period (Table 4.7, Figures 4.3 and 4.4). In most years, a handful of counties had two activities at distinctive levels while all other counties had either none or at most one activity present at levels much higher than the average. While the number of counties with distinctive tourism activities was constant over time, the location of these counties led to intriguing geographic distributions. In 1974 counties that had distinctive levels of establishments that arrange passenger transportation were found near Houston, in East Texas, and in the Texas Panhandle. By 1982 counties where the arrangement of passenger transportation was a distinctive activity had spread to include a cluster of counties adjacent to Houston, five counties adjacent to Dallas–Ft Worth, and a group of counties near Midland and Odessa in West Texas. By contrast, two other parts of the tourism production system, trailer parks and campsites for transients and hotels, tourist courts, and motels, were less frequently distinctive activities in 1982 when compared with 1974 (Table 4.8). As an example of this pattern, in 1974 counties with a distinctively high number of hotels were found along the upper Rio Grande, on the central Gulf Coast, in coastal East Texas, and in scattered rural counties. By 1982 only the scattered rural counties still had a distinctively large number of hotels.

Turning to the description of the tourism production system based on standardized entropy values (Table 4.9, Figures 4.5 and 4.6), over time the number of counties without any tourism declined and the number of counties

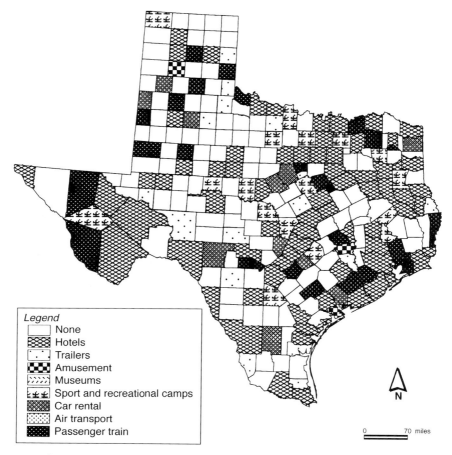

Figure 4.1 Dominant tourism establishments, 1974
Sources: Roehl and Fesenmaier (1988a and 1988b)

with the most highly diversified tourism production systems increased. Geographically, in 1974 Houston, Dallas, Ft. Worth, San Antonio, and most other metropolitan areas had diversified tourism economies. Counties with undiversified tourism production systems were located in East Texas, along the coast, in the Hill Country, in counties adjacent to the Big Bend area of the Rio Grande, and in scattered rural areas. By 1982 a number of counties in East Texas and around Big Bend had developed more diversified tourism economies and were no longer found in the least diversified category. Diversification had also spread to suburban counties surrounding major metropolitan areas, coastal counties, and counties in East Texas and the Texas Hill Country.

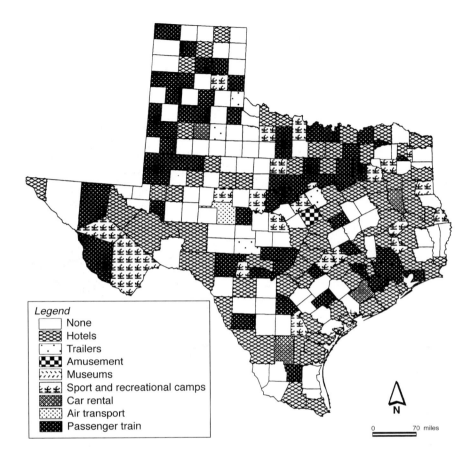

Figure 4.2 Dominant tourism establishments, 1982
Sources: Roehl and Fesenmaier (1988a and 1988b)

To summarize the results of this study:

- By 1982 travel and tourism establishments had spread into counties where they did not exist in 1974.
- Dominant functions appeared to change during this period.
- The evolution of the tourism production system in Texas counties seemed to reflect two different types of destination areas; natural resource areas like East Texas and the Texas Hill Country, and metropolitan areas specializing in urban tourism activities.
- 'Urban' tourism activities, such as hotels, air transportation, museums, and the arrangement of passenger transportation diffused to lower-order

69

urban centers and to suburban counties adjacent to major metropolitan areas.

* As time passed, counties were more likely to have travel and tourism establishments representing several different activities.

Table 4.7 Number of distinctive[a] tourism activities in Texas counties, 1974–82

Number per county	Year					
	1974	*1976*	*1978*	*1980*	*1982*	*Total*
zero	169	165	174	172	168	848
one	78	80	74	78	81	391
two	7	9	6	4	5	31
Total	254	254	254	254	254	1,270

Note:

a Distinctive is defined as those activities in a county at levels more than two standard deviations above the mean for all Texas counties.

Table 4.8 Type of distinctive tourism activities in Texas counties, 1974–82

Economic activity	Year					
	1974	*1976*	*1978*	*1980*	*1982*	*Total*
Air Transportation (451+458)	4	4	6	6	14	34
Arrangement of passenger transportation (4722)	16	12	13	25	23	89
Hotels (7011)	14	17	16	10	5	62
Sport and recreational camps (7032)	11	15	11	15	20	72
Trailer parks and campsites (7033)	13	13	9	7	3	45
Passenger car rental (7512)	16	15	16	12	12	71
Amusement parks (7996)	4	4	3	2	4	17
Museums, etc. (84)	0	0	0	1	0	1
Air transportation and museums (451+458, 84)	4	4	4	4	4	20
Other combinations of activity	3	5	2	0	1	11
None[a]	169	165	174	172	168	848
Total	254	254	254	254	254	1,270

Note:

a These counties either did not contain any travel and tourism establishments, or none of the types of travel and tourism establishments occurred at levels greater than two standard deviations above the mean.

In their second study, Roehl and Fesenmaier (1988b) examined change in the tourism production system in Texas coastal counties (twenty-two counties adjacent to the Gulf of Mexico). In addition to dominant functions, distinctive functions, and functional specialization, this study also examined the relationship between growth in total establishments and growth in the number of establishments that were part of the tourism production system. In ten of the coastal counties, the rate of growth in the number of tourism production system establishments exceeded the rate of growth for all types of establishments during the study period 1974–82. Most of these counties were located near Harris County (Houston). Perhaps the most interesting finding from this study had to do with the role of Houston. In 1974 Harris County accounted for 48.4 per cent of all coastal tourism establishments and 60.9 per cent of all coastal establishments. By 1982 Harris County accounted for 59.5 per cent of the coast's tourism establishments and 65.1 per cent of all the coast's establishments. Associated with Harris County's increased importance, most other counties' share of tourism establishments declined. This pattern was particularly notable in Cameron and Nueces Counties, both major tourism destinations. Cameron County's share of coastal tourism declined from 9.6 per cent to 7.4 per cent of the total, while Nueces County's share of total coastal establishments declined from 10.7 per cent to 7.2 per cent of the total.

However, the general tendency over the study period was for the tourism production system in the area to become more diversified. In 1974 counties on the Texas coast were less diversified than the average for the entire state. By 1982 coastal counties tended to display a much higher level of diversification in their tourism production systems than was found in the state as a whole. Interestingly, this occurred even though the state as a whole became much more diversified over this time period. Statewide, this diversification appeared to represent both the growth of tourism (in number of establishments) as well as the changing visitor demands for additional complementary services. Diversification occurred in counties with large tourism sectors as well as in counties with smaller tourism sectors.

The changes observed in the tourism production systems on the Texas coast seemed to emphasize the importance of location and infrastructure for tourism development. Even though less developed counties experienced growth throughout this period, by 1982 their relative share of tourism economic activity had declined compared to 1974. Areas such as Harris County (Houston) appeared to benefit from its high levels of population growth, as well as its dominant position in infrastructure and accessibility, capturing a larger share of tourism economic activity throughout this period.

This analysis also offers some support for theories of the destination life cycle (Butler 1980). For example, during this time period, trailer parks and campgrounds were replaced by hotels and motels as the dominant activity in a number of the coastal counties. This is the type of change one might expect

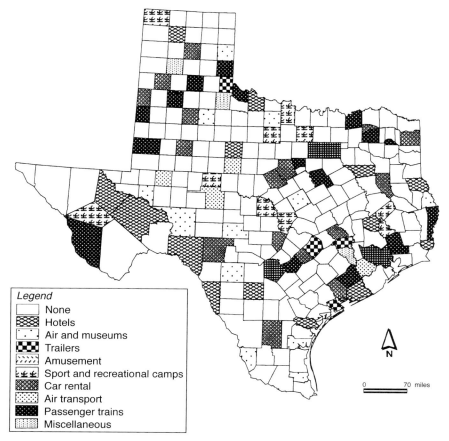

Legend
- None
- Hotels
- Air and museums
- Trailers
- Amusement
- Sport and recreational camps
- Car rental
- Air transport
- Passenger trains
- Miscellaneous

Figure 4.3 Distinctive travel and tourism functions, 1974
Sources: Roehl and Fesenmaier (1988a and 1988b)

as an area matures as a destination. At an earlier stage of development, tourism lodging is provided by trailer parks and campgrounds. As the area matures, the number of visitors increases and these visitors are likely to require more standardized and familiar forms of lodging, such as that provided by hotels and motels. Furthermore, this change from trailer parks and campgrounds to hotels and motels also suggests that local entrepreneurs are being supplemented by firms from outside the area, given the higher levels of investment capital required by hotel and motel development.

While limited by the use of number of firms rather than, for example, number of employees, payroll, or revenue, this pair of studies provides a good example of the type of investigation possible when defining the tourism production system as a collection of specific industries. It also illustrates some

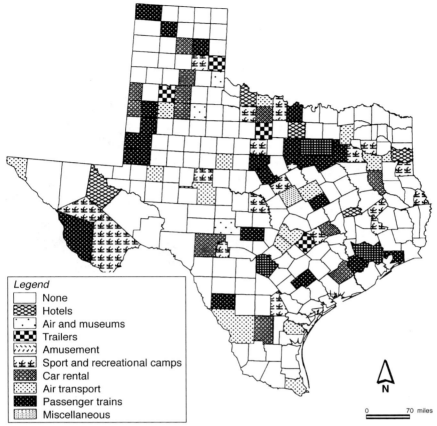

Figure 4.4 Distinctive travel and tourism functions, 1982
Sources: Roehl and Fesenmaier (1988a and 1988b)

of the problems inherent in using this approach to study the tourism production system. For example, by using US SIC 7512, *passenger car rental and leasing, without drivers*, Roehl and Fesenmaier may have contaminated their measurement of the tourism production system by including an activity that is not part of tourism – passenger car leasing.

In the final section to this chapter, some concluding comments will be offered as well as a number of recommendations for future research.

Summary, conclusions, and the future

If tourism is not an industry, in the usual meaning of the term, then what are our options for studying the economic geography of tourism? The satellite

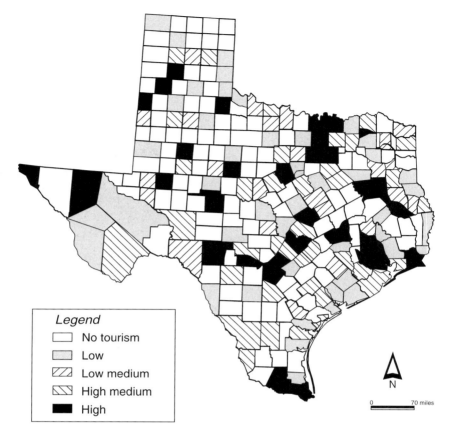

Figure 4.5 Tourism diversification, 1974
Sources: Roehl and Fesenmaier (1988a and 1988b)

Table 4.9 Diversification of the tourism production system, as measured by an entropy coefficient, in Texas counties, 1974–82

Entropy coefficient	*Year*					
	1974	*1976*	*1978*	*1980*	*1982*	*Total*
Less than 0.024	46	45	39	46	38	214
Between 0.024 and 0.353	29	33	29	30	39	160
Between 0.354 and 0.516	37	40	50	38	44	209
Greater than 0.516	32	33	36	42	47	190
No tourism	110	103	100	98	86	497
Total	254	254	254	254	254	1,270

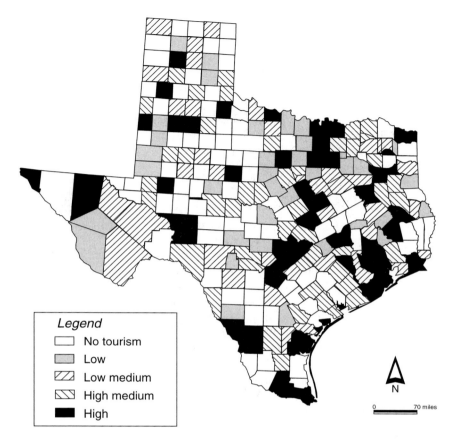

Figure 4.6 Tourism diversification, 1982
Sources: Roehl and Fesenmaier (1988a and 1988b)

account approach is one option, and is discussed by Smith in Chapter 3 of this book. Another approach is to define the tourism production system using an existing, standard, system of industrial classification. That is the approach discussed in this chapter.

In taking this approach, this chapter has attempted to identify some of the issues surrounding the definition and classification of industrial structure. Existing systems of classification have been discussed, with an emphasis placed on the US SIC codes and the new North American standard industrial classification system. In particular, while many users of economic data apparently are not aware of the NAICS (Boettcher 1996), it will create the opportunity for the improved study of tourism geography. The study of services in general, and tourism in particular, will benefit from the added detail the NAICS will provide.

One of the contributions this chapter makes is to provide a preliminary identification of the parts of the tourism production system using NAICS definitions. However, it is important to recognize that one advantage of defining the tourism production system using existing industrial classifications is that it encourages multiple definitions of tourism. Depending on the context, anything from a broad to a narrow tourism production system can be defined. Or multiple tourism production systems can be defined and then some form of sensitivity analysis used to examine the relationship between conclusions about the tourism industry and different definitions of tourism. This approach could make an important contribution in helping researchers discriminate between robust conclusions about tourism and definition-specific conclusions about tourism.

Defining tourism production systems using the NAICS will also facilitate cross-national studies of tourism. This common set of definitions will encourage the analysis of tourism in all three NAFTA countries – the United States, Mexico, and Canada. This will open opportunities for studies of national policy and its impact on tourism.

Finally, using existing systems of industrial classification may also help reintegrate tourism geography into the mainstream of economic geography. Defining tourism with the NAICS, for example, will encourage tourism researchers to view tourism within the context of the broader economy. This may stimulate increased communications between scholars interested in tourism and those working in the more traditional areas of economic geography.

Part B

THE DEMAND-SIDE

5

THE DETERMINANTS OF TOURISM DEMAND

A theoretical perspective

Muzaffer Uysal

Introduction

In recent years a considerable number of empirical and review studies of the determinants of tourism demand have been reported in tourism and related journals. This chapter is intended to provide a theoretical overview of the determinants of demand from a spatial perspective. Although a brief discussion of the supply-side of tourism is inevitable, this chapter focuses on the demand-side of tourism. More specifically, it examines the interaction of demand and supply, and reviews the types of demand measures and determinants.

According to Johnson and Thomas (1992), tourism demand analysis is of interest from several points of view. First, public policymakers need to examine trends in and determinants of demand. Measures of tourism demand can be used to assess the contribution of the tourism industry to the economic welfare of the local economy as a whole and to provide a guide to the use and efficient allocation of resources. Second, management has a strong interest in tourism demand–supply interaction. Marketing decisions and strategic planning of tourism provisions require knowledge of factors affecting destination choice and type of trips, and forecasts of tourism flows in the short and long term. Therefore, one of the purposes of tourism demand studies is to improve the ability to forecast (Witt and Witt 1995) and understand travel behavior.

Spatial interaction between demand and supply

A number of scholars have proposed models of the tourism system (Gunn 1994; Leiper 1979; Mill and Morrison 1985). In its simplest form, the tourism system consists of an origin and a destination. On the one hand, an origin represents the demand-side of tourism, the region or country generating the visitors. A destination, on the other hand, refers to the supply-side of tourism that may have certain attractiveness powers. The tourist and tourism attractions are the central aspects of the system. The transportation and information (marketing) components are seen as 'linkages' which enable the tourist to make decisions concerning where to go, how long to stay, and what to do. These linkages, however, also enable the industry through promotion, product development, and pricing strategies to affect directly the decisions of prospective customers (Fesenmaier and Uysal 1990). The interaction between the two is reciprocal and affects the intensity of demand and travel flows. Figure 5.1 depicts the elements of demand–supply interaction.

Pearce (1995) reviews different types of tourism models demonstrating the nature of interaction between demand and supply. For example, the early explicit models of tourist systems tended to focus on the linkage or travel component, differentiating between recreation travel and pleasure travel, and demand and supply; whereas, the origin–destination models are based on the assumption that most places are, in varying degrees, origins and destinations and serve both as receiving and generating places. Structural models, by contrast, focus on the dominant role of metropolitan countries as generating places and also as bases of air carriers that can effectively and selectively control the international links between the market and a destination. These models suggest that the relationships between the two places occur as a function of the technological and economic superiority of the travel-generating places. Evolutionary models of tourism stress change and examine tourist movements from both the perspective of the evolution of tourist movements and the development of tourism structures. Destinations go through different phases of development (Doxey 1975; Butler 1980; Martin and Uysal 1990; Debbage 1990). Thus, the interaction between the market and destination will change over time in terms of the types of visitor attracted and their behavioral characteristics (Plog 1974).

Conversely, Mansfeld (1990) discusses two basic streams that conceptualize tourist flows in the literature: (a) travel flows as a function of demand–supply interaction (functional approach) and (b) travel flows as a result of political and economic prosperity/superiority. The latter approach argues that travel flows occur 'from the desire of affluent classes in metropolitan countries to travel to dependent and deprived countries' (see, for example, Lea 1988; English 1986). The implicit assumption of this approach is that the flow between an origin and destination(s) is unidirectional as a function of the 'political and economic superiority' of the generating place. This approach is considered radical in its research orientation and does not take into account the behavioral and functional or evolutionary aspects of tourists. However, there is enough evidence to suggest that a significant volume of international visitation and tourism may still be attributable to this theoretical framework (see English 1986; S. Britton 1982). Regardless of the emphasis and approach

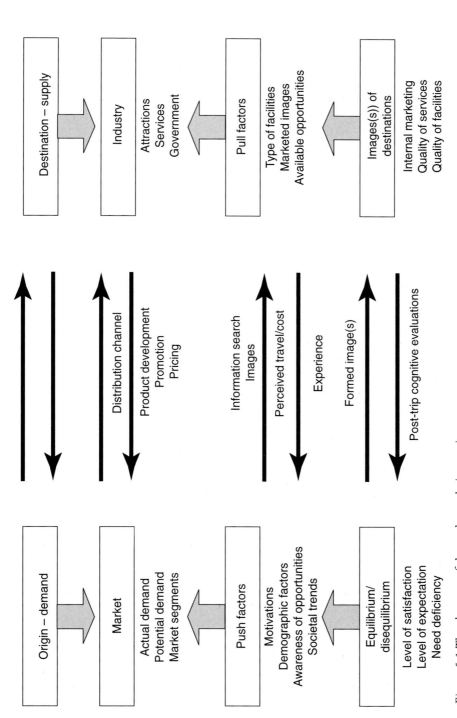

Figure 5.1 The elements of demand–supply interaction
Source: Adapted from Fesenmaier and Uysal (1990)

in each flow model, the essence of the interaction between demand and supply remains a key to the production of tourism goods and services.

According to Rugg (1973), a traveler does not derive utility from possessing or consuming travel destinations; rather, the traveler derives utility from being in the particular destination for some period of time.[1] The very existence of tourism depends on the availability of resources at the destination. The resources that attract tourists are numerous, varied and limited in numbers, in distribution and in degree of development and the extent that they are known to the tourist market (Pearce 1995). On the market side, producers of transport, accommodation, catering, and entertainment services are involved with travel-marketing intermediaries such as tour operators and travel agents. On the supply-side, leisure and recreational activities at destinations are the concern of the different types of tourism supplier including local and state agencies, private business owners, tourism destination organizations, and the providers of infrastructure, and supporting services of tourism. Thus, the supply-side of tourism can be divided into three elements: tourism-oriented products, resident-oriented products, and background tourism elements (Jafari 1982 and 1983). Tourism-oriented products include accommodations, food service, transportation, travel agencies and tour operators, recreation and entertainment, and other travel-trade services. As tourists extend their stay at destination sites, they may increase their use of resident-oriented products which include hospitals, bookstores, barber shops, and so forth. When utilizing such products, tourists also are exposed to or experience background tourism elements such as natural, sociocultural, and human-built attractions that frequently constitute their main reasons for travel. These elements usually collectively produce the ultimate tourism experience and can be examined simultaneously in the same context (Pyo *et al.* 1991). Therefore, the elements of tourism supply resources are not mutually exclusive, but rather are complementary in nature.

People travel or participate in leisure activities because they are 'pushed or pulled' by the forces of motivations and destination attributes. Push factors are considered to be the socio-psychological constructs of the tourists and their environment that predispose the individual to travel or to participate in leisure activities, thus influencing demand. Pull factors, on the other hand, are those that emerge as a result of the attractiveness of a destination and are thought to help establish the chosen destination.[2] However, in order for a destination or site attribute to meaningfully respond to demand or reinforce push factors, it must be perceived and valued (Uysal and Hagan 1993). An important factor affecting this relationship between motivations (push factors) and destination (pull factors) attributes is the notion of accessibility of the sites, the reasons for travel, the level of information about the site, and the destination preferred by the tourist (Gunn 1994; S. Smith 1983). The interaction between demand and supply is essential for the vacation and leisure experience to take place.

The functional framework to approaching tourist space is commonly used to study the nature of tourist systems and travel flows (Gunn 1994; Leiper 1990c; Mansfeld 1990). The functional tourist space is where the tourist (representing demand) travels to the attraction (the tourist product) offered by the tourist industry (the supplier) using the transport medium, which provides the essential linkage between demand and supply areas. The tourism system also reflects the linkage between the needs of the tourist and the ability of the destination to fulfill these needs. Disequilibrium in an individual's cultural–social–psychological needs can be a primary motivation for travel (Crompton 1979). Crompton suggests that individuals live in a social–psychological equilibrium which may become unbalanced over time. This can occur during a period of routinized and repetitive action, such as at work or in the home environment. The need for change, relaxation, or escape from a perceived mundane environment results in psychological disequilibrium. In this sense, it is important that the vacation experience fulfills the individual's expectations. As Ryan (1991) points out, attempts to understand and explain tourism demand without reference to motivations or social change can only yield at best incomplete forecasts of tourism movements between places[3] (see Gartner 1993; Mansfeld 1990; Dann 1996b).

Tourists also form expectations of a destination based upon advertising and promotional campaigns that, in turn, may influence demand for tourism destinations. The quality of the service and the quality of the facility also directly affect the quality of vacation experiences and, thus, the level of future demand. Further, the level of satisfaction that the tourist feels is also dependent upon the ability of the destination to deliver the type of experience which it has marketed (Fesenmaier and Uysal 1990; Ryan 1995).

Murphy (1985) argues that the kind of interaction that will take place in space, according to the functional approach, depends on three principles of spatial interaction: complementarity, transferability and intervening opportunity. The complementarity principle refers to the unilateral, as opposed to bilateral, tourist movements where the latter pattern enhances the former by creating interchangeable origin–destination pairs. The transferability principle refers to accessibility within certain time and cost constraints. Moreover, these constraints affect the intervening opportunities or destinations, suggesting that a high degree of comparable destinations will create substitutions and influence destination selections. However, Mansfeld (1990) points out that such constraints are not the only players in shaping travel decisions. Travel is sometimes perceived as part of a broader tourist adventure and this can explain the choice of a destination farther away than one nearby even though both destinations possess the same set of attributes. Furthermore, the cost constraint does not necessarily lead to shorter trips because distance has become gradually less related to cost in today's travel business. Linkages between the market place and destination are now increasingly governed and managed by information systems relating to both computerized reservation services and communication systems. These technological components of tourism distribution systems have helped facilitate the flow of information at a faster pace and made business transactions instantaneous, resulting in heightened efficiencies, reduced costs of travel and a more effective use of time and resources.

Space is thus responsible for the unequal (or equal) spatial distribution of visitation. This imperfect factor, immobility as a function of distribution of tourism supply resources, creates the comparative advantage of a particular location (Hoover and Giarratani 1984). In this respect, location must be considered as a very special type of differentiation worthy of separate treatment (S. Smith 1995). Unlike style or quality, location is strictly quantifiable in its effects on a destination's demand curve and desired pricing behavior (Bowes 1978). Spatial competition among tourist destinations revolves around maintaining repeat visitors and intensifying their use per visit in terms of number of nights and expenditures; converting their one-time visitors into repeat visitors; converting deferred demand into effective demand; and monitoring changes in potential demand until it is ripe for conversion into effective demand. Effective demand can be actual repeat visitors or one-time visitors (Wahab et al. 1976), while potential demand reflects at least an adequate awareness (or willingness) to visit the destination (Boniface and Cooper 1994). Deferred demand reflects at least some adequate ability to go to the destination. However, due to the lack of awareness of the destination, lack of willingness to visit, or lack of facilities at the destination(s), demand is yet to be realized (Pearce 1989; Lavery 1974).

Measures of demand for tourism

In economic terms, demand can be defined as the quantity of a commodity or service that a community is willing and able to buy during a given time period (Archer 1980). Actual demand represents the quantity of goods and services that consumers require at a given time. Analyzing these demands in tourism involves studying the reasons behind the development and the intensity of tourism flows between destinations (e.g. countries), and levels of participation in on-site tourism activities. Therefore, the determinants of demand and measures of demand for tourism should reflect the scale of analysis (international tourism, domestic tourism, or on-site tourism demand) in terms of the measurement and operationalization of variables. Demand for tourism is measured in different ways. The most commonly employed measures for demand include:

- number of visitor arrivals or number of participants;
- tourism expenditures or receipts;
- length of stay or tourist nights spent at the destination site;
- travel propensity indices.

The measurement of visitor volume in the form of tourist arrivals and the amount of money spent by tourists on goods and services (receipts or expenditures) have dominated tourism demand models at the international level. However, the number of trips taken in a given time period and length of stay (days or nights spent) as measures of demand are commonly used at the national and local levels. For example, Uysal *et al.* (1988) used length of stay for visitation to national parks to explain demand. Length of stay as a measure of demand was operationalized as the number of cross-country skiing (ski-touring) days at the site, and regressed on such variables as direct cost of skiing, distance traveled, number of previous skiing trips, and selected situational characteristics of the site. Mak *et al.* (1977) also developed a behavioral model using simultaneous equations to analyze the determinants of actual length of stay and per capita daily expenditures for mainland US visitors in Hawaii. Empirical results indicated that length of stay does affect the average daily expenditure per person and vice versa. Results also showed that higher income visitors stay longer and spend more per day than visitors with low incomes.

Resort destinations are also interested in the length of stay of tourists, since the night's stay is the basic commodity purchased by tourists. Such a measure for tourism demand might be superior to one reflecting actual tourist numbers since it accounts for changes in the average length of stay, though it does not allow for changes in average daily expenditures (Crouch 1994a).

Sheldon (1993b) examined issues relating to the measurement and forecasting of international tourist expenditures and tourism arrivals. The results showed that these alternative measures of demand fluctuate differently, and that the accuracy of the forecast differs depending on the country. This observation is also consistent with the results previously reported by Uysal and O'Leary (1986). In their study, two alternative measures of demand – number of arrivals and tourist expenditures – were compared with respect to different country destinations. They reported that tourist expenditures as a measure of tourism demand performed better than the use of arrivals. The findings on this issue suggest that alternative measures for demand should be employed to increase the accuracy and reliability of demand studies. For example, Arbel and Ravid (1985) used visits rather than arrivals as a demand measure to derive demand functions for recreation and other tourist facilities in parks. In their study, the demand functions revealed variation with time (number of visits over time) rather than across populations.

Yoon and Shafer (1996), by contrast, explored price-quantity relationships within the various submarkets of tourism demand, signifying the importance of multiple destination trips in demand measures. As Debbage (1991) suggests, in order to estimate accurately travel demand, multiple destination trips need to be considered as well as single destination trips regardless of the type of measures selected for demand. For example, Lue *et al.* (1996) provide an excellent example of multidestination travel behavior. They conceptualized multidestination travel as a constrained choice process in which individuals evaluate travel alternatives as a bundle of attributes. It is hypothesized that variation of the combination of attributes associated with secondary destinations would influence the likelihood of making multidestination trips. The findings of the study indicated that preference for a destination was enhanced by inclusion of a combination of destinations.

Demand for tourism is also expressed in terms of travel propensity. For example, the region/country potential generation index is used to assess the relative capability of a region or country to generate trips (Hudman and Davis 1994; O'Leary *et al.* 1993). The assumption here is that population is an important factor to consider in the generation of trips. The population can be thought of as a potential pool from which tourists are generated. If we assume that there is a ceiling on tourism demand per person (Edwards 1985), then the size of the population determines the limits on tourism generation from a given market place. Schmidhauser (1975 and 1976) differentiates two types of travel propensity: net and gross. Net travel propensity refers to the proportion of the total population, or a particular group in the population, who have made at least one trip away from home in the period in question. Gross travel propensity, by contrast, is defined as the total number of trips taken in relation to the population under study. The ratio (gross travel propensity/net travel propensity) between the two gives travel frequency. Such indexes appear to be a useful complement to the traditional absolute travel demand and net flow figures (Uysal *et al.* 1995). However, the selection of measures for demand are constrained by the time horizon, availability of data, and the scale of analysis.[4]

Tourism demand determinants

Since the early 1960s numerous empirical studies have been undertaken to develop an understanding of the determinant factors of demand for tourism (Crouch 1994b). Classical economic theory states that the quantity of a good or service demanded is a function of its price, the income of the consumer, the prices of related goods, and personal preferences of consumer tastes. Thus, demand theory directs the inclusion of such variables as prices and income.[5]

Even though income and price related factors are likely to play a major role in determining demand for tourism, as embedded in economic theory, the number of potential demand determinants in the tourism literature is almost unlimited. The possible factors that are likely to influence demand may be divided into three types of demand determinants: (a) exogenous (business environment) determinants, (b) social–psychological determinants, and (c) economic determinants (Figure 5.2).

The exogenous determinants of tourism demand usually deal with general business trends, advancements in use of technology and communication, shifts

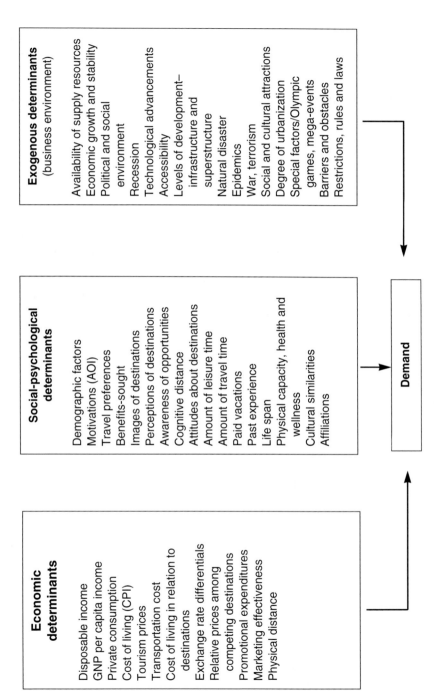

Economic determinants

Disposable income
GNP per capita income
Private consumption
Cost of living (CPI)
Tourism prices
Transportation cost
Cost of living in relation to destinations
Exchange rate differentials
Relative prices among competing destinations
Promotional expenditures
Marketing effectiveness
Physical distance

Social-psychological determinants

Demographic factors
Motivations (AOI)
Travel preferences
Benefits-sought
Images of destinations
Perceptions of destinations
Awareness of opportunities
Cognitive distance
Attitudes about destinations
Amount of leisure time
Amount of travel time
Paid vacations
Past experience
Life span
Physical capacity, health and wellness
Cultural similarities
Affiliations

Exogenous determinants (business environment)

Availability of supply resources
Economic growth and stability
Political and social environment
Recession
Technological advancements
Accessibility
Levels of development– infrastructure and superstructure
Natural disaster
Epidemics
War, terrorism
Social and cultural attractions
Degree of urbanization
Special factors/Olympic games, mega-events
Barriers and obstacles
Restrictions, rules and laws

Demand

Figure 5.2 Determinants of tourism demand

Note: This exhibit is not intended to provide an exhaustive list of factors, but rather to give examples of factors that are likely to affect demand.

in the market place, economic growth, the political and social environments of destinations, and the development and availability of supply resources.

Social–psychological determinants are not easy to measure and are not easily incorporated into economic demand models. These factors are usually examined in relation to travel decision making and destination selections (Sirakaya *et al.* 1996)[6] and are vital in any understanding of the market and its demand for tourism.

Economic determinants of demand are the easiest to measure and are commonly employed in tourism demand studies. These determinants of demand need to be more fully augmented with the social and psychological determinants of tourism in order to more fully understand travel behavior.

Vanhove (1980) reviewed several demand studies and found that the five most commonly occurring economic variables affecting tourism demand are: income levels, population, relative prices, exchange rate, and travel costs. In addition to such variables, Vanhove also mentions as important demand determinants: promotion, common language, accommodation capacity, carrying capacity, and the degree of urbanization. Earlier studies by Armstrong (1972) and Schmulmeister (1979) also provided a similar list of variables.

Because of the difficulty involved in relating the volume of demand to so many variables at once, researchers usually focus on the most influential variables to delineate the determinants of demand for tourism. The following section focuses mainly on the commonly used economic determinants of demand, along with some marketing and supply-related factors.

Income

Income is the variable most commonly used to explain and determine tourism demand. Changes in consumer income can cause changes in the demand for goods and services. An increase in real income provides consumers with greater purchasing power. In tourism demand functions, the real per capita income of tourist generating countries is commonly used (Archer 1980; Uysal and Crompton 1984; Carey 1991; Crouch 1994b; Moshirian 1993). The form of income, however, should be adjusted to accommodate different tourism contexts (Kim 1996). For example, if holiday visits or visits to friends and relatives are under consideration, then the appropriate form of the variable is private consumption or personal disposable income (Witt 1980; Quayson and Var 1982; Summary 1987). If attention focuses on business visits, then a more general income variable should be used (Witt and Witt 1995).

Price

Price is an important factor in determining demand, because it reflects another dimension of purchasing power. However, the role of price is a more complex construct than income in the case of international tourism.

According to Martin and Witt (1987), there are two elements of price: the cost of travel to the destination and the cost of living for the tourist in the destination. These elements of price exert influence on the intensity of travel flows, and the amount of demand for a given place. The cost of living is represented by the price of tourist goods and services in the destination, and is sometimes separated by the effect of exchange rate variations on purchasing power. Another aspect of the price variable is the price of other substitute or complementary products (Pyo *et al.* 1991). Such a practice in demand analysis allows for the impact of competing and intervening destinations by specifying the tourists' cost of living variable relative to a set of alternative destinations.

Due to the unavailability of price data and the difficulty in obtaining uniform price data, a consumer price index (CPI) is usually regarded as a reasonable proxy for the cost of tourism in national and international demand models. However, in the context of recreation demand at the local level, entry fees and charges may be more appropriate. The use of a CPI is justified on the grounds of convenience (sometimes the data are readily available) and the argument that tourists' spending is spread over a wide range of the economy and so may approximate the general average consumer spending patterns, or that at least the CPI will track tourism prices closely (Morley 1994). This argument is also supported by Crouch (1992). Ideally, to measure the responsiveness of tourism demand to the level of relative prices, a tourist services index is needed. Such an index would specifically measure the relative prices of tourist services. Consumer price indices, which measure changes in prices across a broad spectrum of consumption, do not necessarily reflect changes in travel-related prices (Uysal 1985). Although the United States Travel Data Center (USTDC) and Statistics Canada publish a monthly travel price index, a similar type of index is not available for all tourist generating places and receiving places. In those cases where no such indices are available, the consumer price index has to be used as a proxy for the price of tourism services. Traditionally, economic theory assumes that quantity of demand for a product declines as the price of the product increases. In other words, tourists are likely to react when there is a change in the ratio between prices at an origin in relation to price at the destination.

Exchange rate

Tourists are also expected to be affected by the price of foreign currency. If a currency devalues in a foreign country, international tourism becomes 'less expensive' and results in increased travel flows to that country. Conversely, an increase in the value of a country's currency will make international tourism 'more expensive' and cause decreased travel in that country. Therefore, exchange rates can have a significant effect on the extent of international travel. The usual justification for including an exchange rate variable in demand functions is that consumers are more aware of exchange rates than prices in the destination (Witt and Witt 1995). This variable usually appears in addition to a consumer price proxy or as a sole representation of a tourist's living cost (Quayson and Var 1982; Arbel and Ravid 1985).

The Economist Intelligence Unit (EIU) (1975) identified the impacts of an unfavorable change in exchange rates to include: (a) less travel abroad, (b) travel to different locations, (c) a reduction in expenditure and/or length of stay, (d) changes in the mode or time of travel, and (e) a reduction in spending by business travelers. Martin and Witt (1987), however, argue that the exchange rate on its own is not an acceptable proxy, but the consumer price index adjusted by the exchange rate is a reasonable proxy for the cost of tourism. In other words, exchange rates may be used both as a separate independent variable and as a price deflator.

Trade volume

Gray (1970) delineates international tourism demand as a part of a larger invisible international trade system although tourism expenditures and receipts are included in the balance of payments of national accounts. The amount of trade volume between countries may encourage not only business travel but also subsequent trips for pleasure. It is, therefore, logical that trade volume should be included and analyzed in a macroeconomic study of tourism-demand generation. So far, the relationship between international trade and tourism generation has received little attention in terms of its role as a determinant of tourism demand. However, some recent studies have incorporated trade volume as a possible determinant for international tourism on the assumption that business activities between two places or countries would also encourage subsequent pleasure travel (Lee 1994; Kim 1996).

Marketing variables

Promotional expenditures are also suggested to influence demand for tourism, yet are not generally included in international tourism-demand models (Carey 1991). Generally, promotional expenditures spent by tourist offices and organizations are expected to play a role in determining the level of tourism demand since these activities are destination specific and are more likely to influence tourist flows to the destination concerned (Witt and Martin 1987). However, a major problem regarding the inclusion of a marketing variable relates to difficulties in obtaining relevant data. In addition, the impact of advertising on tourism demand may be distributed over time.

Supply factors

It has been suggested that tourism flows are directly influenced by the nature of supply resources, and by the way they adapt to demand (Vellas and Becherel 1995). Some empirical studies support this argument. For example, Kim (1996) reported that, for lodging services in Korea, the supply-side is accounted for by the change of prices responding to fluctuating volume and quantity demanded. The market price of lodging services is found to be related to demand and supply influences simultaneously. Price is generally regarded as exogenous in previous tourism-demand research. In fact, price is not predetermined and changes simultaneously with demand. Most demand studies assume an elasticity of tourism supply, though there actually exists a certain rigidity of supply (Witt 1980). A good example is the fixed capacity of lodging (e.g. hotel rooms, availability of meeting space) over the short-run. Thus, the market price of a given supply resource is related to demand and supply influences simultaneously, making price an endogenous factor as well (Kim and Uysal 1997). These empirical findings confirm the common belief that there exists a certain amount of rigidity in the tourism market, and it takes time for supply to adjust to demand. Nevertheless, a few studies have incorporated supply variables (e.g. bed-nights available, hotel rooms, size of lodging) as constraints into demand models.

Use of dummy variables

A review of the literature on tourism demand studies reveals that the inclusion of dummy variables for a variety of purposes is also common practice (Crouch 1994b). Dummy variables are introduced to account for the effect of special events that may have a transitory influence on demand. In cross-sectional studies, dummy variables are occasionally incorporated to facilitate the estimation of different demand coefficients by country of origin or destination (Kim 1996). In time-series studies involving time periods shorter than one year, dummy variables are used to allow for the effect of seasonality. These demand determinants also cover special factors such as political disturbances, exchange rate controls, recessions, mega-events, and language similarities between origins and destinations (Johnson and Ashworth 1990). Uysal and Crompton (1984) also used a dummy variable to portray social/economic instability in the destination country. In her study, Summary (1987) incorporated a dummy variable designed to measure the impact of a border closure upon wanderlust travelers.

Concluding comments

Tourism is a spatial phenomenon and tourism experiences take place in physical space. Better understanding the interaction between physical resources and demand is of vital importance if we are to understand the intensity and determinants of demand for tourism destinations. The examination of tourism-demand determinants necessitates the inclusion of more than conventional demand variables but also socio-psychological and dummy variables that more effectively capture the demand for tourist experiences. Furthermore, macro issues such as economic growth and prosperity, favorable trade and travel policies, and advancements in technology and communication collectively can also affect travel flows.

Projections by researchers (Edgell 1993; Qu and Zhang 1996) indicate that international tourism will become a more important part of the world economy as tourism continues to grow more rapidly than other areas of the economy. This expected growth in tourism will generate increased demand for leisure travel. More leisure, increased income, better education, paid vacations, modern transportation, and advancements in technology and communication will all continue to contribute to the changing lifestyles which make consumers increasingly mobile, travel oriented, and self-enriched. The tourism industry is in a state of flux. The globalization of tourism has resulted in more diverse travelers with different preferences, expectations and needs. At the same time, the world-wide movement toward democratization has made some locations more accessible than ever before, while creating new destinations. Demand for such new destinations is likely to be affected by very local factors ranging from destination development, safety and security, and air access to political and economic stability. The result of such globalization may also force destinations (countries) to specialize in providing particular types of tourism goods (nature-based tourism, history and heritage) to remain competitive and increase demand for tourism (Urry 1990).[7]

There is often a positive correlation between a country's economic growth and the level of tourism demand. Provided that significant economic disparities do not exist, and certain barriers to travel are eased and removed, accessible places that are rich in 'tourism resources' are likely to attract more visitors. The actual volume of visitation is likely to be determined by several key, interrelated, demand determinants, including various socio-economic variables, psychographic variables, destination attributes, barriers to travel, intervening destinations, preferences, quality of service, accessibility, promotion, marketing effectiveness, perception of the destination and the level of destination development. Because of the difficulty involved in relating the volume of demand to so many variables at once, researchers typically attempt to isolate what they consider to be the most influential group of variables and relate the demand volume to changes in these key variables. It is very important that future researchers appropriately operationalize the determinants of demand and select an appropriate demand estimation method for the study place in question.

Demand is the fundamental measure of any area's success in attracting and serving visitors. The need to enhance the understanding of the determinants of tourism demand at all levels – international, national, regional and local – is of immense value to those who are promoting destinations, monitoring tourism activities and formulating strategies in tourism.

From a marketing point of view, an effective response to market trends and changes would require better knowledge of travelers and their characteristics. The quality and availability of tourism-supply-resources are a critical element in meeting the needs of the ever-changing and growing tourism market. As D. Taylor (1980) suggests, if the goods and services required by the visitor are known, it is possible to list their availability in an area and determine how well the supply matches the demand. As a marketing tool, a supply–demand interaction system allows an area to be carefully matched with present and potential visitors. Tourism suppliers need to be knowledgeable about the expectations and needs of different tourist types so that appropriate goods and services can be delivered to sustain and increase demand for their destinations (D. Taylor 1996). Therefore, the physical space in which the tourism experiences take place plays a crucial role in configuring the tourism demand and supply interaction. The development and modification of tourism resources and products, in accordance to changes in consumer preferences and expectations, can help ensure continued success in developing the tourism product.

For this reason, changes in consumer tastes and economic indicators need to be constantly monitored so that the extent to which these changes have influenced demand can also be assessed over time. Such efforts can not only provide timely and accurate information but also form a sound data base for decision makers and policymakers in the tourism sector.

Notes

1 The relative amount of vacation time spent in a place within a destination has received attention from several researchers in analyzing spatial distribution of visitors with respect to places or destinations (Pearce and Elliott 1983; Uysal and McDonald 1989; Debbage 1991; Murphy 1992; Uysal et al. 1994; Oppermann 1992 and 1993). The amount of time spent in a place in a given destination during a total trip or vacation is established by the use of the Trip Index as suggested by Pearce and Elliott (1983). The Trip Index has been utilized to identify main, gateways, short-trip destinations, secondary and stop-over destinations, and segment markets for promotional and planning purposes. The index compares the days (or nights) spent at the destination to the total length of the trip or vacation. It is determined by dividing the days (nights) spent at the destination by the days (nights) of the entire trip and multiplying the result by 100. A Trip Index of 100 indicates that the entire trip is being spent at the one destination, thereby suggesting it is a primary destination; a value of zero, on the other hand, indicates a 'pass-through place' whereby visitors spend no more than a few hours within that place.

2 Husbands (1983) differentiates between the concepts of attraction versus

attractiveness. He argued that attractiveness emanates from a destination attribute, however attraction has more to do with the actual levels or patterns of visitation. He further added that actual visitation does not only depend on destination attractiveness (spatial variation), but also on other variables such as marketed image and advertising.

3 Ryan (1991), in his book *Recreation Tourism: A Social Science Perspective*, provides an excellent discussion on the nature and importance of social and psychological determinants of demand. For further information on these issues, see chapter 2, pp. 14–34.

4 For more information on statistical measurements and definitions of tourist arrivals versus tourist persons, length of stay, and other related issues, see Strong (1992) pp. 735–45; and S. Smith (1995), chapter 2, pp. 20–39. The discourse in chapter 2 of Smith's book regarding the desire for common terms and definitions by the WTO is excellent and provides an adequate discussion of the difficulties associated with properly defining the tourism system and its components. The development of a standard industrial classification of tourism activities (SICTA) is also described.

5 The most common statistical method used to estimate such demand variables is Ordinary Least Squares (OLS) Multiple Regression (Uysal and Crompton 1985; Sheldon and Var 1985; Martin and Witt 1989; Summary 1987; Crouch 1994a). The use of the OLS Multiple Regression is based on the assumption that a demand measure (dependent variable) has a cause and effect relationship with one or more explanatory (demand determinants) variables (Var and Lee 1993). Under certain assumptions, the method of OLS has some very attractive statistical properties which have made it one of the most powerful and popular methods of regression analysis. Kerlinger (1986) points out that multiple regression analysis is an efficient and powerful precision tool with complex interactions between independent variables and the dependent variable(s), and this helps to explain the presumed phenomenon represented by the dependent variable.

In terms of the functional form of the demand models, there appears to be an almost universal agreement that the multiplicative (log linear) form is superior to the additive (linear) form (Johnson and Ashworth 1990). The multiplicative model often fits the demand data better and conveniently provides constant demand elasticities (Morley 1994; Kim 1996). Such elasticities of demand determinants are then used to formulate policies and examine how consumers respond to changes in demand determinants. However, a recent study by Qui and Zhang (1995) found that the estimation of functional forms (log-linear and linear forms) could show variation from country to country. For example, in tourist arrivals equations, the log-linear functional form performed sightly better than the linear functional form for the United States and Japan, but linear forms were slightly better for the United Kingdom, France and West Germany. In tourist expenditure equations, the log-linear functional form performed better only for France, but linear forms were better for the United States, the United Kingdom, and West Germany.

6 For further discussion on the importance of social and psychological determinants of tourism in decision making, see Crotts and van Raaij (1994); Louviere and Timmermans (1990); Um and Crompton (1990); and Woodside and Lysonski (1989).

7 Urry (1990) in his book *The Tourist Gaze: Leisure and Travel in Contemporary Societies* discusses new challenges and opportunities due to the changing economics of the tourist industry, see chapter 3, pp. 41–65; Ioannides (1995) stresses the importance of such economic issues as tourism production system, entrepreneurship and labor pool. Also, see Go and Pine (1995). This book examines the recent expansion of multinational hotel companies and the challenges they face in the international arena, and illustrates strategic responses employed by different companies. Chapter 3, 'Appraising the global competitive environment' (pp. 50–93), discusses in particular how the new information technologies are changing the nature of demand and creating new forms of competition in the international market place.

Part C

NEO-FORDISM AND FLEXIBILITY: A SECTORAL APPROACH

6

NEO-FORDISM AND FLEXIBLE SPECIALIZATION IN THE TRAVEL INDUSTRY

Dissecting the polyglot

Dimitri Ioannides and Keith G. Debbage

Introduction

In the last decade, a number of academics have argued that leisure-related activities are becoming increasingly commodified reflecting the emergence of a global 'culture of consumption' (S. Britton 1991; Boorstein 1964; Cohen 1995; Featherstone 1991; Pretes 1995; Urry 1995; Zukin 1988). Certain authors (Feifer 1985; Mullins 1991; Urry 1988 and 1990) contend that tourism has itself undergone significant changes mirroring a broader process of societal reorganization, particularly the shift towards post-modernism. In the leisure and travel industry, a focal point has been the growing emphasis on the creation and marketing of tourist experiences through place commodification. Prototypes of these landscapes include but are not limited to: amusement and theme parks; casinos; spectacle events such as the Olympics; and festival marketplaces like Boston's Faneuil Hall, London's Covent Garden, and Baltimore's Inner Harbor.

According to S. Britton, places are marketed as desirable products for tourists 'not necessarily as ends in themselves, but because visits to them, and the seeking of anticipated signs and symbols, are a vehicle for experiences which are to be collected, consumed, and compared' (1991: 465). Consequently, the travel industry is responsible for not only marketing and selling the means to an end (an airplane seat or hotel room) but the end itself, the travel experience or what Urry (1990) calls 'the tourist gaze.'

Some post-modern theorists argue that the commodification of experience as an intangible element of capitalistic accumulation has brought the emergence of 'symbolic or cultural capital' (Bourdieu 1984; Harvey 1989a; Jameson 1984; Zukin 1988). In reference to the tourist gaze, cultural capital broadly involves the consumption and collection of touristic experiences to demonstrate social tastes, lifestyle choices, and socio-economic status. Under the logic of this post-modern culture of consumption, the constant search for novelty and alternative experiences is emphasized, as the consumer becomes more discerning and sophisticated. As Poon (1993) argues, in recent years travelers have become increasingly dissatisfied with inflexible, standardized, mass tourism products and, instead, have begun seeking custom travel products tailored to their own particular requirements. As a result, in the travel industry, a premium has been increasingly placed on providing the post-modern citizen or Feiffer's (1985) 'post-tourist' with an unending range of novel travel experiences.

These developments reflect contradictions because while they complement they *also* subvert the alleged trend towards individuation of leisure,[1] with the corresponding differentiating of lifestyles and experiences (Rojek 1985). By personalizing and differentiating travel and leisure products through niche marketing, design variations, and advertising, the travel industry is able to 'camouflage' the industrialized and mass produced nature of such products (S. Britton 1991).

In this chapter we argue that the tension between individual identity and the larger-scale capitalistic mechanisms delivering the tourist gaze to the consumer have produced a unique, complex, and inchoate polyglot of travel-related production processes, which places a premium on increasingly flexible forms of accumulation. Some tourism researchers (S. Britton 1991; Urry 1990 and 1995) have integrated the flexible production literature into explanations of changing patterns of travel and tourism, but the flexible production thesis has been almost entirely discussed in relationship to the manufacturing industry (Amin 1989; Dicken 1986; Massey 1984; Scott 1993) and, albeit to a lesser extent, the producer services (Christopherson 1993; Daniels 1993; Noyelle and Stanback 1984; Urry 1987; P. Wood 1991b). A closer examination of specific product supply arrangements in the travel industry can help shed new light on ways in which innovative flexible production technologies and management-related techniques have shaped and created novel post-modern tourist landscapes.

The chapter highlights how the travel industry resolves the paradox of providing individually-tailored experiences within the context of the larger capitalistic machine that acts to commodify the same experience. Our approach is far from comprehensive. Instead, we restrict our attention to one theme: the shift from mass-packaged tourism to more flexible travel arrangements reflects broader changes towards so-called 'neo-Fordist' (Malecki 1995)[2] and 'post-modern' consumption, and the effects of this shift are felt at all levels of the travel and tourism industry. It is, of course, impractical to cover all the points of a broader agenda for flexibility and the travel industry polyglot, and so we avoid the thorny issue of the manner in which consumption can shape production and the convoluted point of whether a tourism production chain that coherently shapes tourist experiences really exists. Nor do we examine all sectors of the vast travel industry complex. Instead, we focus narrowly on those parts that seem most susceptible to neo-Fordist, flexible production approaches (e.g. airlines, cruise lines, hotels, travel agents and tour operators). Indeed, we have chosen Bull and Church's (1994) approach of investigating fairly coherent industrial groupings with clear products as opposed to exploring the hard-to-conceptualize, amorphous tourism sector as a whole. We also do not thoroughly explore the experiential aspects of tourism, but focus on the ways in which the flexible production framework in some sectors of the travel industry has helped redefine the tourist gaze. Thus, our chapter is a productionist reading of the theoretical and management literature as it relates to the travel industry polyglot.

We now turn to a more detailed discussion of Fordist and flexible-based production techniques and examine whether or not these ideas apply to the travel industry polyglot thesis.

Flexible production

Several observers (Harvey 1989a; Piore and Sabel 1984; Storper and Walker 1989) maintain that since the early 1970s capitalist society and, particularly, industrial structure have undergone paradigmatic shifts in production philosophies. These theorized changes in industrial production are formally labeled the shift from Fordism to neo-Fordism or flexible specialization (Table 6.1). This flexibility thesis has become a focal area of economic geography research in recent years (Malmberg 1994; Yeung 1994) despite certain critics' warnings (e.g. Gertler 1988; Jessop 1992; Sayer 1989) that flexible special-ization has been 'collapsed unquestionably into the notion of post-Fordism' (Malecki 1995: 11).

Deriving its name from Henry Ford's automated assembly line, Fordism describes the dominant industrial production form for much of this century (Table 6.2a) (Yeung 1994). Fordism focuses on the mass production of stan-dardized goods which are, in turn, assembled in large stocks. The emphasis is on economies of scale, translating into major fixed capital investments and industrial concentration through vertical and, to a lesser degree, horizontal integration (Harvey 1989a). The principle labor characteristic under Fordism is that each worker performs a single, highly specialized function along the assembly line.

Since the 1970s, many have argued that the Fordist mode of production has been in 'crisis' (Sayer 1989: 669) (Table 6.2b). In particular, information technologies and robotics have enabled firms to shift from the rigidity of the Fordist-based assembly line toward flexible production techniques and flexible labor markets. Flexibility distinguishes neo-Fordist production systems from earlier Fordist practices (Mullins 1991) because, under this new system, the focus has switched to economies of scope (or network-based economies) and high levels of product differentiation through small batch

Table 6.1 Glossary of key terms

Pre-Fordism	Also known as the craft or artisanal stage of industrial production (c. before 1920). The main characteristic of pre-Fordism is small-scale production of goods targeting reduced markets through non-institutionalized, independently-run 'cottage' industries. Production is labor intensive and uses minimal technology.
Fordism	The dominant mode of industrial production in western societies (c. 1920–1970s). The main characteristic of Fordism is the rigidity associated with mass production of standardized goods along an assembly line. Production is resource driven and labor is characterized by a high degree of job specialization (Table 6.2a).
Post-Fordism	Also known as flexible accumulation, this mode of industrial production relies increasingly on flexible production techniques and flexible labor processes (c. post-1970s). Production is demand-driven, relying on technology (e.g. robotics) (Table 6.2b).
Numerical flexibility	The ability of firms to change levels of labor input according to varying output levels.
Functional flexibility	The ability of workers to move between an operation's different functional tasks because of changes in the nature of the work load.
Horizontal integration	Expansion of a firm's market share within its existing product line. (For example, in the travel industry: airline network expansion through acquisition of other air carriers.)
Vertical integration	Expansion of a firm through a particular industry's various product lines. (For example, in the travel industry: airlines acquiring hotel chains or tour operators controlling retail travel outlets.)
Vertical disintegration	Control by main enterprise of only the final product while peripheral activities are sub-contracted to other firms.
Outsourcing/subcontracting	The process of distancing non-strategic functions (e.g. peripheral labor tasks) to other firms. (For example, in the travel industry: hotels subcontract laundry operations or specialized kitchen activities.)
Product differentiation	The marketing strategy of adding new products, whether this involves minor design changes or radically new products. This results in an expansion in

Table 6.1 continued

	the breadth of product variants available at a particular time. (For example, in the travel industry: a number of hotel-based holding companies have a wide variety of hotel chains, each emphasizing a particular market niche appeal.)
Strategic network alliances	Co-operative business arrangements between competing firms to solidify market positions and extend the provision of seamless customer service to a broader market area.
Economies of scale	Productivity increases arising from increasing the scale of operation. In general the attainment of lower costs of production at large levels of output.
Economies of scope	Also known as network economies. Maximizing market dominance and efficiencies through network-based externalities to achieve a sustainable competitive advantage. (For example, airline hub networks.)

production of various specialized commodities targeting a multitude of market niches (Debbage 1992). Rather than relying on large stocks of homogeneous (just-in-case) goods, neo-Fordist industries have the flexibility to alter significantly the delivery of goods and services in response to varying consumer demand, through 'just-in-time' (JITs) inventories (Table 6.2b).[3]

A hallmark of flexible production is vertical disintegration, meaning a central 'enterprise controls the final product and the key technology' (Coffey and Bailly 1992: 858) while non-strategic functions are sub-contracted to other firms. The externalization of certain production functions gives firms the advantage of various cost savings plus external economies of scale. Instead of directly producing its own goods, for example, the Benetton clothing company functions as a central marketing enterprise, controlling a network of independent small specialist producers (Coffey and Bailly 1992). Similarly, the US-based Nike company subcontracts its entire athletic shoe production process out to overseas independent firms (Donaghu and Barff 1990).

An analogous situation exists in parts of the travel industry (Table 6.2b). Responding to an increasingly sophisticated clientele, many hotel holding companies have created niche brands, each catering to one market segment (Go and Pine 1995), while diverging from the direct ownership of actual properties through the increasingly ubiquitous more flexible management contract or franchise agreement (Ioannides 1995). The availability of CRS (computer reservation systems) technologies has also created opportunities for the creation of small and medium-sized firm networks in the travel industry (Buhalis 1993). Concurrently, subcontracting has become increasingly common, as evidenced by outside contractors obtaining maintenance

Table 6.2 From Fordism to flexible specialization – theorized transformation in manufacturing and tourism

Fordist production in manufacturing (1920s to 1970s)	Fordist production in tourism (1950s to 1990s)
	a
The production process	
Economies of scale	Economies of scale
Mass production of homogeneous goods	Mass, standardized and rigidly packaged holidays
Dedicated (assembly line), inflexible machinery	Packaged tours, charter flights
Uniformity and standardization	Narrow range of standardized travel products
Large buffer (just-in-case) stocks and inventory	Holding holidays 'just-in-case'
Production is resource driven	Tour industry determines quality and type of product
Industrial concentration (vertical and to a lesser degree horizontal integration)	Industrial concentration (horizontal and to lesser extent vertical integration)
Labour practices	
Functional and numerical inflexibility – single task performance by specialized worker	Low labor (functional) flexibility
	High labor turnover, labor is seasonal, low wages
Little on-the-job training	Mostly unskilled labor force
The consumption process	
Mass consumption	Mass tourists
Customers are inexperienced, motivated by price	Tourists are psychocentrics (inexperienced, predictable), sun-lust, motivated by price

Table 6.2 continued

Post-Fordist production in manufacturing (post 1970s)	b 'Post-Fordism' and 'flexibility' in tourism (1990s to future)
	The production process
Economies of scope Small-batch production of a variety of customized product types Market niching Information technologies and robotics	Economies of scale *and* scope Emergence of specialized operators, tailor-made holidays Market niching System of information technologies (SIT) (CRS technology, teleconferencing, videotext, videobrochures, satellite printers, etc.), front and back-office automation, Internet, World Wide Web
No stocks, just-in-time inventories Production is demand driven Vertical disintegration, subcontracting of non-strategic functions	Custom-designed, flexible holidays Tourists determine product type Horizontal integration, subcontracting (e.g. the hotel industry externalizes laundry operations or specialized kitchen activities)
Inter-firm strategic alliances	Adoption of regionally based, integrated, computer information systems and strategic network alliances in the airline industry
	Labour practices
Functional and numerical flexibility	Functionally flexible (skilled) year-round employees flanked by peripheral, numerically flexible, unskilled workers
Employment security for core workers, no security for temporary workers	
	The consumption process
Individualized consumption Very experienced consumers Greater volatility of consumer preferences Increased preference for non-mass forms of production and consumption	Independent tourists Experienced, independent, flexible (sun-plus) travelers Fewer repeat visits Demand for 'green tourism' or other alternative forms (e.g. ecotourism)

Sources: Based on Coffey and Bailly (1992); Harvey (1989a); Poon (1993); Urry (1995).

agreements with major airlines (*Aviation Week and Space Technology* 1993) or hotels externalizing their laundry services (McDermott and Martinez 1989).

Beyond production techniques, flexibility in neo-Fordism extends to labor practices. Firms commonly practice *numerical flexibility*, whereby they adjust the aggregate quantity of labor in response to fluctuations in demand for individual products (Coffey and Bailly 1992; S. Wood 1989). This means many companies' labor structures have become increasingly bifurcated, with a few highly skilled, well-paid, permanently employed core workers enjoying high job security surrounded by a larger group of occasional, low wage, unskilled laborers. Moreover, many post-Fordist companies practice *functional flexibility*, where firms use the same worker for a variety of tasks, indicating a firm's capacity to match constantly its labor force's skills to changing production circumstances. Several travel and tourism labor markets have long experienced both numerical and functional flexibility. Thus, these can hardly be considered innovative practices in the industry. For example, because of seasonality, many sectors of the travel industry have traditionally depended on temporary or part-time, unskilled or semi-skilled workers (Shaw and Williams 1994). These peripheral employees flank a core of year-round, functionally flexible workers responsible for a variety of tasks, especially in small independently owned lodging establishments in traditional resorts with marked seasonality (Urry 1990).

Such trends indicate that parts of the travel and tourism supply system experience varying degrees of flexibility in terms of production and labor practices. A key question is to which sectors of the travel and tourism economy can the theorized shift from Fordism to flexible production be most easily applied? Also, given the travel industry's service orientation, is it appropriate to think of the production of mass-market-oriented tour packages as analogous to the Fordist practice of punching out automobiles along an assembly line (Malmberg 1994)? Perhaps most critically, given that vertical disintegration and the externalization of non-strategic activities appear to be a central hallmark of a flexible production system, what analogies, if any, exist in the production of travel-related services? Finally, can a clear chronological transition in production innovation be articulated or is the amorphous travel industry a polyglot of varying production processes and management strategies that defies categorization?

The travel industry polyglot

Though the theorized transition from Fordism to neo-Fordism refers primarily to manufacturing (Malmberg 1994), some authors have discussed its relevance to producer services (Daniels 1993; Coffey and Bailly 1992). However, most of the flexible production literature continues to underplay the role of consumer services, especially the travel-related economic sectors.

Some travel and tourism researchers (Urry 1990; Mullins 1991; Page 1995;

Poon 1989) believe the transition from Fordism to flexible specialization has some application to the travel and tourism industry. For example, Urry (1988) argues that in response to the alleged advent of post-modern culture, tourism has itself become a post-modern phenomenon. According to Urry, if 'the British seaside resort was the quintessential form of tourism in industrial society [then] with more recent changes in the direction of a post-industrial society, there has been the emergence of what has been journalistically characterized as "post-tourism"' (1988: 35–36). Thus, Urry believes the old 'Fordist' seaside holiday camps of the 1930s, 1940s, and 1950s have transformed into post-Fordist places of 'freedom' (Urry 1990: 14) emphasizing individuality over any perceived notion of mass markets.

Moreover, a considerable body of literature focuses on how the process of deindustrialization reconfigures cities from places of production to consumption (S. Britton 1991; Featherstone 1991; Mullins 1991; Urry 1990). Often, the very sites representing the cities' productionist or mercantile past (e.g. industrial plants, warehouses, market places) have themselves become objects of the tourist gaze. Yet, Mullins (1991) cautions that much of the discussion concerning the post-modern city is based on imprecise interpretations. He recommends that to comprehend the post-modern city, the transition from Fordist to flexible production and consumption must first be understood.

In contrast to this more theoretical work, Poon (1993) examines the 'real world' strategic imperatives facing the travel industry in particular. Poon argues that the transition from Fordism to flexible production in manufacturing is mirrored by equivalent shifts in travel and tourism and that major technological innovations (particularly the advent of CRSs), accompanied by new trends in consumer behavior, allow rigid mass-oriented, standardized package tourism to be replaced by more flexible travel forms emphasizing individuality and autonomy.

While these contentions are intriguing, they demonstrate weak links with the broader concepts articulated in the flexible production literature. Importantly, they fail to establish how the travel industry as a production system extends the flexibility framework. A particularly problematic aspect of most such analyses (e.g. Poon's) is their longitudinal perspective, assuming a clear chronological break between Fordist and neo-Fordist tourism activities. In fact, these analyses demonstrate similar problems to the ones characterizing the work of many post-Fordist theorists, namely the tendency to view the shift from Fordism to flexible specialization in manufacturing as unidirectional, and their insistence that one has to choose either Fordism *or* post-Fordism without 'recognition of the intermeshing of different production systems in both time and space' (Yeung 1994: 465).

As with manufacturing and producer services, it is important to realize that the emergence of flexible production processes in some parts of the travel industry did not suddenly emerge with the appearance of terms like post-Fordism in the literature. 'Rather, such processes have existed to varying degrees for centuries within both manufacturing and service production activities [including tourism]; however, these processes have become increasingly present during the past decade' (Coffey and Bailly 1992: 866). In a sector as amorphous as the travel industry, with so many permeable boundaries and so many diverse linkage arrangements to exploit, a polyglot of coexisting multiple incarnations has evolved, displaying varying traits of flexibility.

True, some elements of the travel industry demonstrate the markedly neo-Fordist characteristics of enhanced flexibility, particularly based on information technologies, plus externalization of services (e.g. certain new entrant airlines and specialist tour operators). Still other components fit the mold of rigid, standardized, mass-market, Fordist production processes, though even these sectors frequently attempt to leverage flexibility through brand segmentation (e.g. large hotel chains, cruise lines). Interestingly, a substantial part of the travel industry (particularly at the point of contact with the customer) remains distinctly pre-Fordist or artisanal in its make-up (i.e. based on small or medium-scale, family-run operations) (e.g. country inns and bed and breakfast establishments). This sector also displays high degrees of flexibility (especially in terms of labor practices). Moreover, CRSs and other information technologies (e.g. access to the Internet and the World Wide Web) offer the opportunity for these pre-Fordist establishments to create small firm networks and, thus, become increasingly flexible. The emphasis on flexibility (albeit to varying degrees) in many parts of the travel industry polyglot partly reflects an attempt to reconcile what amounts to an impersonal and large-scale capitalistic production process with broader post-modern trends that encourage the individuation of leisure (S. Britton 1991).

The chapter now turns to a more explicit and detailed discussion of the travel industry polyglot and the differing levels of flexibility inherent to pre-Fordist, Fordist, and neo-Fordist production methods.

Pre-Fordist travel industry

The 'pre-Fordist', 'craft', or 'artisanal' travel industry (Fayos-Solá et al. 1994: 10) describes the sector's substantial non-institutionalized element that existed long before the Fordist revolution of the 1960s. The pre-Fordist travel industry generally includes lodging and eating establishments, plus businesses such as nightclubs and souvenir stores, most of which are small to medium scale, family-owned and operated. It also applies to a substantial part of the travel agency business, which continues to be dominated by a large number of independently run, small 'mom-and-pop' businesses (Ioannides 1994).

Labor characteristics

While the travel industry's pre-Fordist sector tends to be small-sized and independently owned, it is also pervasive in terms of employment generation. For example, in Greece and France, most travel-related enterprises (i.e. hotels and restaurants) are family owned, employing fewer than ten persons (Leontidou 1988). In Britain, where the average hotel has twenty-five rooms, the majority of hotel rooms continue to be provided in family-run businesses like guest houses or bed and breakfast establishments (Shaw and Williams 1994). Despite a slight proportional increase in the share of rooms controlled by major hotel chains and marketing consortia, 70 per cent remain in the hands of independent entrepreneurs (Urry 1990).

In the pre-Fordist travel industry, the small, independently owned establishments are frequently characterized by weak management skills, rely heavily on family labor and low levels of capital investment, and are economically marginal (Shaw and Williams 1994). Not surprisingly, therefore, ownership turnover is high. In Britain, approximately 25 per cent of small tourist accommodation businesses have a survival span of two years (Urry 1990).

In a pre-Fordist context, the travel industry often is a side-business for people otherwise formally engaged in economic sectors like agriculture (e.g. renting out rooms on the farm). Ironically, the small size and informal nature of the majority of pre-Fordist-based tourism establishments mean labor flexibility (particularly functional flexibility) is common. As Urry acknowledges, 'the flexible use of labour, is something that has characterized many tourist-related services for some decades' (1990: 78). It is not unusual for the owner of the small establishment with a limited workforce to work long hours and double up as receptionist, chef, and/or waiter, not to mention 'assisting with washing up, when and where the need arises' (McDermott and Martinez 1989: 192). Because of seasonality, plus occasional demand peaks (e.g. public holidays), numerical flexibility is also exercised regularly in artisanal tourism establishments. During busy seasons the proprietor may hire casual workers, such as family members (e.g. children on summer vacation) and/or friends. Most likely, these occasional workers are the least functionally flexible and are normally hired to fulfill a single low-skill task.

The role of information technology

In pre-Fordist travel establishments, levels of technology are minimal or non-existent. While many lodging or eating establishments are gradually revolutionizing their kitchen technology (e.g. through dishwashers and microwaves), a large proportion of artisanal travel establishments do not use any information technologies. In a recent survey of cottage-letting industries in the UK, 40.5 per cent of respondents claimed they did not use personal computers (Mutch 1995). Given the increased affordability of personal computers in recent years, this lack of interest in information technologies can be attributed largely to inexperience. Moreover, many owners of artisanal tourism establishments may fear that computers will eliminate the 'personal touch' between themselves and consumers (Mutch 1995). Indeed, the unique service-oriented features of many parts of the travel industry may mean certain segments of the industry are particularly resistant to certain techno-logical innovations.

A major problem for small travel enterprises is their inability to gain easy access to CRS technologies or global distribution systems (GDS) (see Chapter 7 in this book). Milne and Nowosielski noted that in South Pacific microstates 'the costs of joining computer reservation systems and the skilled labour required to operate them may place them out of the hands of smaller, locally owned concerns' (1997: 7). This phenomenon is not unique to less developed countries. For instance, CRSs do not list 90 per cent of Montreal's small accommodation establishments (Pohlman 1994). Likewise, a pilot survey of special-interest, tour-operating firms (niche specialists) in the United States (see Chapter 8 in this book), identified only 16 per cent of responding companies with links to CRSs.

Nevertheless, technological innovations which may yet enable enhanced flexibility for the pre-Fordist component of the travel business include the Internet and World Wide Web (WWW). As the costs of computerization and going 'on line' continue to descend, small niche market specialists (tour organizers and lodging establishments among others) gain the opportunity to market their products directly to consumers, possibly eliminating the services of traditional travel intermediaries and, more importantly, reducing the control of CRSs. Already in the United States many small travel-related companies, including a significant number of specialist tour operators, have set up their own WWW pages (Walle 1996).

Fordist travel industry

Although the ensembles of pre-Fordist travel-related establishments are highly visible, the industry's fortunes are largely dictated by a few powerful travel companies (e.g. airlines, transnational hotels, cruise companies, and tour operators). Since the 1960s, these firms, all based in the principal tourist origins of the industrialized world, have emphasized globalization strategies, reflecting characteristics which have become increasingly common in both the manufacturing plus the producer and consumer service sectors (Daniels 1993; Dicken 1992; Knox and Agnew 1994; Ioannides 1995). Referring to the globalization of services, Daniels believes the main reason for their expan-sion across national borders has been the search for new markets 'in order to sustain growth and profit expectations [because] deregulation, accelerating technological innovation and the rapid emergence of new competitors have undermined long-standing market boundaries' (Daniels 1993: 43). Reflecting these trends, major hotel companies have been especially keen to expand their operations internationally (Go and Pine 1995; Ioannides 1995), while airlines (e.g. British Airways) have also adopted globalization tactics, albeit to a lesser extent (Debbage 1994).

Linked to these globalization strategies, major travel-related corporations pursue market share and increased market concentration, resulting in a highly oligopolistic industrial structure that can easily manipulate origin–destination flows on a grand scale (Ioannides 1995). A dominant trait of these large-scale, travel-related corporations is their ability to leverage significant economies of scale through the high sales volume of affordable, standardized products, especially tour packages. Given the emphasis on the mass production of travel products, and the analogies which can be drawn from the production assembly line in manufacturing, these elements of the industry are labeled Fordist.

Economies of scale

'Economies of scale' in the travel industry refers to the process whereby direct costs continue to decrease proportionately with volume because suppliers can give high discounts for larger bookings (Pearce 1989). Transnational hotel chains, for instance, have major marketing and management advantages over their smaller competitors. These chains are better able to fund staff training and enjoy far more efficient production methods (Dunning and McQueen 1982b). Also, construction costs decrease with increasing plant size which explains the trend towards building larger hotels. For instance, while in 1948 less than 45 per cent of US rooms were in hotels of more than 100 rooms, by 1985 this figure had risen to approximately 62 per cent (Go and Pine 1995).

Along similar lines, major tour operators (e.g. Thomson and Kuoni) rely on volume bookings by mass tourists because of significant cost advantages in packaging the various travel components into single products. Such benefits 'look unassailable in the next decade . . . [because] . . . the gap between the full cost of independent travel paying published tariffs, compared with inclusive tour prices, is likely to widen rather than narrow' (Middleton 1991: 192).

Industrial concentration and globalization

Behind the Fordist and frequently oligopolistic structure of sectors like the airline, tour operating, and hotel industry has been the process of horizontal (and to a lesser extent vertical) integration. A select few transnational hotel companies now dominate the hospitality industry, especially through franchise and contract management agreements (Go and Pine 1995). By 1994 Hospitality Franchise Systems (HFS) had acquired the rights to chains like

Ramada Inn, Howard Johnson, and Days Inn and had overtaken Holiday Inn Worldwide to become the world's largest lodging company (Standard and Poor 1995). HFS currently controls close to 400,000 hotel rooms in 3,873 properties worldwide. Moreover, the top four United States based chains (HFS, Holiday Inn Worldwide, Best Western International, and Choice Hotels International) control more than 11,500 properties (approximately 1.25 million rooms) world-wide.

The major tour operators located in Western Europe and North America also exhibit a highly consolidated market structure (see Chapter 8 in this book). In the late 1980s, for instance, out of approximately 670 British travel wholesale companies, just two (Thomson and the International Leisure Group) controlled more than 53 per cent of the market. By 1989 77.5 per cent of inclusive tour holidays originating in the UK were sold by the leading five operators (Sinclair 1991). In the travel agency sector it has been very much the same story. In 1991 the top fifty travel agencies in the United States owned 4,000 retail outlets and were responsible for one third of total agency airline ticket sales. American Express, the largest firm, sold $4.5 billion worth of tickets accounting for 10 per cent of total airline sales volume in the United States (*Travel Weekly* 1992).

Correspondingly, the airline industry has become increasingly oligopolistic since it was deregulated in the United States in 1978. By 1993 the largest US carrier (United Airlines) controlled more than 20 per cent of the domestic market. At the same time, the top five airlines' market share exceeded 80 per cent (Standard and Poor 1994). A similar trend has occurred in European countries, where following moves towards privatization and more recently liberalization, a handful of mega-carriers (e.g. British Airways, Lufthansa) have emerged (Debbage 1994; Ioannides 1995). Though it is unlikely that the effects of air liberalization in Europe will be as abrupt as those of deregulation in the United States, some analysts contend that the region's share of world airline traffic will eventually increase substantially from the 27 per cent it controls today (*Aviation Week and Space Technology* 1997).

The cruise line industry, yet another Fordist representation of the travel industry, also displays increasing market consolidation. In 1992 the top three cruise lines in the United States controlled 47 per cent of the market share while the top six companies accounted for approximately two-thirds of the market (Hobson 1993). By 1996 the three big companies Royal Caribbean Cruise Lines (CCL), Princess Cruises, and Carnival Corporation had increased their market share to 65 per cent (*Travel Weekly* 1996). Analysts predict that by 1999 the market share of these three companies will expand by a further 10 per cent.

While increased market concentration through horizontal integration is a trait of the travel industry's Fordist component, vertical integration (a hallmark of Fordist production in manufacturing) has never been widespread. True, select examples of vertical integration exist, particularly in Europe where major tour-operating firms operate their own charter airlines and/or travel agencies (O'Brien 1990). Similarly, in 1992 the parent company of CCL also owned the Crystal Palace Resort and Casino in the Bahamas (Hobson 1993), while the Canadian Government has more recently allowed the Greyhound Canada bus company to launch a low-cost air service (*Wall Street Journal* 1996a). A more common form of 'quasi-vertical' integration, however, involves the association of travel-related companies with essentially non-travel-based corporations. For example, Bass PLC, the British brewery, controls Holiday Inn Worldwide (Go and Pine 1995; Standard and Poor 1995), while HFS Inc. now accounts for 25 per cent of the United States' residential, real-estate brokerage market having purchased Century 21, ERA, and Caldwell Banker in the mid-1990s (*The New York Times* 1996a) and more recently Avis car rentals. Other examples of quasi-vertical associations in the travel business include the presence of hotel companies in the gaming industry, plus the agreements enabling the customers of credit card or long-distance telephone companies to accrue frequent flyer miles for airline travel.

Nevertheless, more traditional forms of vertical integration in the travel industry have proven problematic. For example, the Allegis umbrella company collapsed in the mid-1980s after United Airlines sold Hertz car rentals, Hilton International, and Westin, two years after first purchasing these companies. A significant reason behind the limited success of vertical integration in the travel industry is that 'the fortunes of airlines, hotels, and tour operators tend to rise together, depending on the overall health of the economy' (R. Britton 1978: 169). Unlike many joint business ventures where one component of the relationship bolsters the other during a crisis, an airline hurt by an economic downturn cannot be supported by its subsidiary hotel chain or tour operator since they would also suffer business losses. Furthermore, the overall complexity of the amorphous travel industry means most firms have recognized the need to focus on their core business. For example, the Manor Care hotel franchise plans to spin off the Choice Hotels International units from its other substantial interests in the health-care industry because of the lack of synergies between the various business units (*The New York Times* 1996a).

Product differentiation and brand super-segmentation

Despite the rigidity associated with the mass production of standardized Fordist travel, trends in some key travel-related sectors of the economy reflect the adoption of competitive strategies aimed at leveraging enhanced flexibility. Some of the most visible evidence of enhanced flexibility in already established, Fordist, travel industry firms exists in the hotel industry where several major chains have developed sophisticated product differentiation and brand super-segmentation strategies, targeting specific market niches (Rogers

1993). Brand super-segmentation strategies are the Fordist hotel companies' response 'to the plurality of today's marketplace [where] each of today's consumer groups has a specific image of the services they want and the prices they are willing to pay' (Witham 1985: 39). Moreover, as Go and Pine (1995) argue, super-segmentation reflects the hotel industry's response to the shortage of suitable sites for expansion. Since in most metropolitan areas the prime sites have already been occupied by hotels, the only way to expand is to launch novel product lines. Thus, if the mid-range market has been saturated in a particular locale, then a hotel can establish itself by creating a product geared to the lower or upper-end of the market.

> Today's contemporary hotels, designed to please almost every taste and income level, are as rich in variety as in location, and quite a departure from an era marked by sameness and complacency, when downtown hotels, highway motels and resorts virtually monopolized the lodging industry.
>
> (Go and Pine 1995: 102)

One of the first companies to introduce a more sophisticated form of product differentiation to the hotel industry was Quality Inns, largely in response to the blurred consumer image that its vastly diverse properties were promoting (Go and Pine 1995). Many other lodging companies have followed suit. The Marriot chain, for example, has shifted from its long-held position at the higher end of the market, by targeting the mid-priced market through its Courtyard, Residence Inn, and Fairfield Inn hotels (Standard and Poor 1995). Similarly, Holiday Inn's Express Hotels cater to budget travelers while the Crowne Plaza Hotels are geared towards the upper end of the market. Moreover, the French based ACCOR company offers a variety of distinct accommodation products. ACCOR's Sofitel caters to the needs of the luxury market while Novotel and Ibis are respectively tailored for the mid-scale and economy markets. Product differentiation does not only occur within a hotel company but also within individual hotel properties. For instance, in some of its properties, Sheraton offers executive floors designed with the needs of the business traveler in mind (Go and Pine 1995).

Super-segmentation has also become commonplace in the cruise-line industry despite the sector's Fordist, mass-market-oriented image. Most major cruise lines now cater to a variety of segments divided along the lines of income, age, or family status. A rapidly expanding segment of the cruise line industry is geared towards speciality markets such as adventure travelers, sports enthusiasts or those who seek an educational experience (Hobson 1993; *CLIA Cruise News* 1997). For example, Silversea Cruises in conjunction with the National Geographic Traveler offers cruises to persons who want to learn more about the destinations they are visiting, while Norwegian Cruise Line has begun operating 'Sports Afloat' theme cruises featuring football, hockey, and other activities.

Despite adopting such flexible bearing strategies, most holding companies still emphasize large-scale industrial organization and an essentially Fordist *modus operandi*. With reference to hotels, for example, brand super-segmentation and the development of multiple chains within the same holding company reflect the lodging industry's attempt to camouflage the mass production of its services in order to cater better to the specific personal needs and the overall trend towards the individuation of travel.

Information technologies

CRSs and GDSs are commonly hosted by large-scale Fordist-based transnational companies (e.g. hotel chains, airlines, and car rental agencies). These technologies are best designed to serve the requirements of business travelers with fixed itineraries since the latter normally demand scheduled services and standardized hotel rooms (Buhalis 1993). Moreover, hotel chains and major airlines have their own computerized systems with direct links to the principal CRSs. This means the best represented tours on a travel agent's CRS monitor are most likely those offered as standardized packages (e.g. cruises or fixed itinerary coach excursions). By contrast, budget travelers seeking least-cost travel alternatives through smaller, more flexible production firms (e.g. charter flights and specialty operators) are unlikely to find much information through a CRS.

Nevertheless, the use of information technologies in the travel industry allows major Fordist-based tour-operating companies enhanced flexibility in organizing and producing travel services. These suppliers practice wider market segmentation by providing an array of tailor-made products at prices competitive with those of organized mass tours. Rather than catering only to traditional niches divided along socio-economic or demographic lines, suppliers economically offer service packages to cluster segments (O'Brien 1990). Other information technology applications enhancing flexibility within the Fordist travel industry include the adoption of information technologies in dealing with resource-consuming tasks (Gamble 1989). Even though the lodging industry is traditionally 'considered a technology laggard' (Go and Pine 1995: 113), many hotel companies have over the last two decades embraced automated accounting and front office systems plus CRSs. In order to reduce operating costs, hotel chains seek to reduce labor intensive functions and have turned to automated check-in/check-out. In the United States, most major downtown hotels now enable their guests to examine their bill and check-out through their room's interactive television set (Gamble 1989).

'Neo-Fordist' travel industry?

Because parts of the travel industry continue to be dominated by a polyglot of pre-Fordist and Fordist elements partially displaying flexible production tendencies, it is inappropriate to discuss a clean break from the standardized Fordist industry of the 1960s and 1970s towards a totally new post-Fordist travel industry firm. While information technology innovations, plus factors like enhanced industrial concentration and airline deregulation, have certainly enabled enhanced flexibility in the travel industry by order of magnitude, such trends surely pale in comparison to those associated with the post-war invention of wide-bodied jets and the related advent of mass tourism.

Nevertheless, given the enhanced flexibility characterizing the production and management of the pre-Fordist (i.e. small-scale enterprises) and Fordist (i.e. large-scale industrial organizations) components of the travel industry, it is appropriate to term this sector 'neo-Fordist'. As opposed to 'post-Fordism', the term 'neo-Fordism' does not imply a radical change from old practices. Rather, in the context of the travel industry, 'neo-Fordism' refers to the inherent contradiction of increased levels of industrial concentration and increasingly large-scale industrial organizations, while simultaneously implementing increasingly sophisticated, brand, super-segmentation strategies and highly flexible travel-based products.

One of the hallmarks of post-Fordist production systems in the manufacturing sector is vertical disintegration or the externalization of peripheral services through sub-contracting (Coffey and Bailly 1992). Such tendencies have resulted in the proliferation of inter-firm, inter-establishment structures where a network of small and medium-sized specialist enterprises supports the activities of the core business. One wonders whether such changes in other sectors of the economy also have relevance in the travel industry?

Outsourcing

Since the externalization of non-strategic services through outsourcing occurs in certain segments of the travel industry, it appears to fit, at least partially, the mold of post-Fordist practices as they apply to manufacturing. Outsourcing of ancillary activities provides numerous advantages, including the generation of external economies (Coffey and Bailly 1992). Companies accrue considerable savings by buying cheaper services (e.g. aircraft maintenance or aircraft leasing) from outside specialist firms, because the latter can generate scale economies. It is not cost effective for a tourism-related firm to keep a permanent, highly skilled staff to deal with specialized services, given that the demand for many of these functions can be both irregular and unpredictable (Bull and Church 1994). Moreover, the labor reduction achieved through outsourcing results in considerable savings when it comes to employee benefit packages.

Lodging establishments now outsource various specialist services. For example, the hotel industry contracts out ancillary activities such as car parking, laundry operations, catering or specialized kitchen activities (e.g. pastry making) (McDermott and Martinez 1989). According to Bull and Church (1994), during the period 1980–87 the number of catering contractors in the United Kingdom rose by 60 per cent while during the same period the number of hotels and catering establishments grew by only 10 per cent. In most instances, only peripheral activities are outsourced. It is unlikely, for instance, for a hotel to hire outside agency cleaners as opposed to their own.

Airlines rely heavily on contract caterers, in part because catering logistics can represent up to 15 per cent of airline costs (*Air Transport World* 1997).[4] The world-wide in-flight catering market grossed $9 billion in 1993 led by LSG Lufthansa Service/SKYCHEFS which served 310,000 meals daily to over 200 carriers at seventy-two different locations. LSG, which employed over 18,000 workers, earned $1.34 billion, representing a 15 per cent market share. Indeed, the top four airline caterers world-wide controlled half the market (*Air Transport World* 1994).

Moreover, in order to save considerable costs, certain US-based carriers (particularly new entrants) have turned to part-time, independent customer service representatives (ICSRs) (*Aviation Week and Space Technology* 1996a). These representatives work out of their own homes rather than a traditional call-center facility, offering overflow and back-up reservation services when needed. ICSRs allow airlines the flexibility to reduce the inefficiencies encountered with the use of regular, full-time employees because of daily or seasonal peaks. These 'virtual' call centers place a premium on flexibility and the externalization of peripheral services.

Similarly, to reduce costs and enhance profitability, many airlines use third-party contractors for maintenance and repair. Prior to US airline deregulation in 1978, the outsourcing of aircraft maintenance barely existed but is expected to cost $30 billion by 2005 (*Wall Street Journal* 1996b). For example, in early 1996 ValuJet used six or more outside private contract maintenance facilities – most of which were not airline operators. However, following the ValuJet crash in May 1996, the Federal Aviation Administration (FAA) has expressed concern about the quality of work performed by some of these outside contractors. In response, the low-cost airline has attempted to entice Air Canada or the AMR Corporation (American Airlines) to take over its heavy maintenance tasks. Furthermore, the structure of maintenance outsourcing actually varies significantly from airline to airline. The AMR Corporation, for example, continues to control maintenance functions relating to precision work (e.g. jet engine overhaul) while it may outsource more non-safety related functions like seat-cover repair.[5]

Along with maintenance operations, airlines increasingly outsource information systems (e.g. the recent reorganization of American Airlines' Sabre CRS as an independent subsidiary with a view towards a partial or total spin-off later) (*Aviation Week and Space Technology* 1996b). Airlines also out-source asset ownership such as the purchase of new aircraft. For example, in 1995 46 per cent of the aircraft flown by the world's airlines were leased to carriers by companies like the GPA group (*Aviation Week and Space Technology* 1996c). Engine leases have also become increasingly common in the 1990s as airlines seek for new ways to unload equipment risk (*Air Transport World* 1997).

Contracting out services has become so common in the airline industry that many carriers have purportedly been turned into 'virtual airlines' whose business focus is to carry passengers and freight. For example, British Airways (BA) is now examining the possibility of outsourcing additional functions such as baggage handling, cargo, or ramp-vehicle maintenance and has considered making its lucrative maintenance division into a separate company. It appears that BA's information technology department and its catering services may also eventually be outsourced. 'Outsourcing is being viewed by many airlines not only as a logical method of reducing costs, but of eliminating expensive spare parts inventories and releasing carriers to focus on their core competencies – transporting passengers and freight' (*Aviation Week and Space Technology* 1996c: 35) or, in other words, focusing on the business and leisure travel function of the airline industry. It is presently unclear whether the 'virtual airline' will one day become a reality. Robert Britton, Managing Director of Corporate Communications at American Airlines, expresses caution by saying that while 'there is genuine opportunity to reduce expense . . . such savings will be worthless if customer service is noticeably degraded'.[6] Moreover, in certain European countries there still remain a number of impediments to outsourcing what have traditionally been regarded as in-house functions. For example, employment regulations in Italy prohibit crew leasing (*Air Transport World* 1997).

Strategic network alliances

If outsourcing is a clear manifestation of post-Fordist production tendencies, then by default another related outcome is the proliferation of inter-establishment networks based on these various sub-contracted arrangements – the so-called strategically-based network alliance. Some travel-related sectors of the economy have realized that the formation of strategic network alliances allows for a more effective pooling of resources, improved marketing coverage, and technology sharing. Already, a number of regionally-based, integrated, computer information reservation management systems have been developed in Europe and North America (Buhalis 1993). These systems provide information (and sometimes reservations) for the components making up a region's entire travel and tourism product. Importantly, they enable small and medium-sized tourism enterprises to amass collective bargaining leverage when dealing with CRS operators and other powerful industry players.

Many hurdles (e.g. limited understanding of technology and reluctance to cooperate with other industry players) have, thus far, prohibited the wide-scale adoption of network alliances between small or medium-scale, pre-Fordist tourist enterprises. However, the airline industry appears to have been more successful in establishing cooperative strategic alliances between major carriers (e.g. KLM and Northwest, Lufthansa and United) (Debbage 1994). If approved, the global alliance between American Airlines and British Airways would mean that the two airlines will cement their stranglehold on the US–UK air traffic, and control 25 per cent of traffic between the United States and Europe (*Air Transport World* 1997). Global alliances between large airlines and small, feeder-based or commuter carriers are also becoming more common. For example, there has been a proliferation of code-sharing agreements or joint marketing ventures between major carriers like Lufthansa and smaller regional-feeder airlines like Air Mauritius. The national carrier of Cyprus has already indicated its intention to form alliances with a major European Airline should the island become a member of the European Union (Wheatcroft 1994). In 1996, there were 389 alliances among 171 airlines, up from 280 alliances among 136 airlines in 1994 (*The New York Times* 1996c).

Conclusions

The travel industry displays a complex and inchoate polyglot of production forms, many attempting, albeit at varying degrees, to achieve enhanced flexibility. Thus, neatly bracketing the amorphous travel industry into purely pre-Fordist, Fordist, or post-Fordist elements is impossible. Similarly, it is not viable to identify clear chronological transitions from one form of the travel industry to another given the amorphous nature of the industry and its related suppliers. Instead, the travel industry can be classified as 'neo-Fordist' since it demonstrates varying degrees of flexibility enhancement within the context of its pre-Fordist establishments and institutions, as well as its dominant Fordist elements.

The evidence presented indicates the flexible production framework can and should be extended to account for the production of consumer-related services like travel. Much like manufacturing or producer services, the travel sector has been significantly affected by technological innovation, particularly the adoption of new information technologies. Importantly, parts of the travel industry exhibit tendencies normally associated with flexible production in manufacturing like the externalization of ancillary activities, the development of inter-firm strategic network alliances, and product differentiation through brand segmentation.

If the shift to post-Fordist consumption and flexible specialization is itself part of a broader post-modern phenomenon, it seems that parts of the travel industry polyglot have embraced the post-modern ethos. Part of the rationale for these structural changes involves allowing consumers a measure of autonomy in shaping their own travel experience (Rojek's (1985) individuation of leisure and Featherstone's (1991) culture of consumption). However, part of the rationale is less theoretically exotic and more pragmatic – travel businesses seek to maintain a competitive advantage (Porter 1985 and 1990) in the marketplace by cutting the cost of doing business (e.g. outsourcing of peripheral businesses).

Disentangling these respective motivations must be left to future scholars wishing to embark on more detailed investigations of the corporate strategies of a specific travel industry supplier, particularly as they relate to adopting flexible production management innovations. A useful starting point may be to examine the competitive strategies of the airline industry and tour operators, largely because both sectors have the ability to exert enormous influence over inter-industry transactions and the geography of origin–destination flows.

On the one hand, airlines can mold travel distribution channels not only through the selection of specific air routes to particular resort destinations but also through sophisticated CRSs like the American Airlines Sabre system. Most airline-based CRSs are heavily used by travel agents, and through travel agent commissions, the airline industry can manipulate the type of advice given by the agents. On the other hand, tour operators occupy a strategic position by virtue of their ability to leverage and negotiate packaged deals of enormous volume between all the principal suppliers (see Chapter 8 in this book). Importantly, both sectors of the travel industry are busily preparing for the highly flexible, neo-Fordist world of the future by developing, among other things, specialist tour-operator subsidiaries and niche markets, and heavily outsourced airlines which may eventually become 'virtual' companies.

A key problem is whether these paradigmatic shifts in production reflect broader shifts in consumer tastes or simply the fashion of practices like outsourcing, strategic alliance networks, or brand-differentiation strategies existing in other economic spheres. For example, Urry (1990 and 1995) argues that post-Fordist consumption is consumption rather than production dominant because travel and tourism services have to be produced and consumed at the same location. While there is a level of 'spatial fixity' in the provision of tourist services, in the sense that the production of these 'services cannot be entirely carried out backstage, away from the gaze of tourists' (Urry 1995: 40), it is unclear whether this is an appropriate rationale for arguing that the production-based relationships in the travel industry are significantly culturally defined. Resolving this 'chicken or egg' paradox may require intelligible answers to the following questions:

- What parts of the travel industry polyglot are most susceptible to the externalization of particular consumer service-related functions, and why?

- What will be the impact of technological changes, like ticketless travel, on the decision by hotels or airlines to externalize peripheral activities? For example, American Airlines currently has a 'back-office' operation with 1,000 employees in Barbados where the flight coupons are sent for processing. The introduction of ticketless travel will gradually eliminate the task and will have a negative impact on the Barbadian economy.
- Does the composition of the labor force in any given travel-related sector of the economy have a direct bearing on externalization strategies?
- What, if any, is the influence of firm size or spatial contingencies on the decision to externalize specific types of travel-related activities?
- How pervasive is the establishment of inter-firm based networks and strategic alliances in the travel industry, and which economic sectors and resort locations are best suited to this form of industrial structure?
- What are the implications of flexible production innovations for governments and institutions in tourist destinations? Specifically, can these destinations, especially in less developed countries, use technology and flexible management bearing strategies to improve their competitiveness in the global market place?

Effectively answering these and other research questions can help develop an understanding of how tourist-resort cycles are connected to broader changes in post-Fordist and post-modern consumption and allow us to determine better if 'tourist resorts typically display all the features of flexible-production agglomerations' (Scott 1993: 265). Finally, it is important to recognize that many parts of the travel industry have a long history of practicing flexible labor practices and many types of processes that have been considered in this chapter did not just suddenly appear when the term 'flexible production' entered the current literature. Still, to understand better management strategies in the travel sector, it is time to appreciate better both the polyglot of flexible production techniques currently practiced in the industry and how this relates to the broader literature on flexible production management practices.

Acknowledgement

A substantial part of this chapter has been reprinted from *Tourism Management* 18,4, Dimitri Ioannides and Keith G. Debbage, 'Post-Fordism and Flexibility: The Travel Industry Polygot', pp. 229–41, 1997, with permission from Elsevier Science Ltd, The Boulevard, Langford Lane, Kidlington OX5 1GB, UK.

Notes

1 According to Rojek (1985), the individuation of leisure is one of four elements characterizing the organization of leisure in modern capitalism.

2 We have chosen Dunford and Benko's (1991) term 'neo-Fordist' as opposed to the more commonly used 'post-Fordist'.

3 While many of the changes in production have been precipitated by changes in consumption, a full examination of changing consumer tastes is beyond the scope of this chapter which focuses on the major travel-related suppliers and the changing methods of production. In the travel industry these changes in consumer demand have become increasingly complex and varied. They include: diminished leisure time and purchasing power, fewer children in families, delayed families, psychographic changes, and an increased demand for more flexible and specialized packages among others (Plog 1991).

4 Food can represent 20 per cent of catering logistics. Other logistics include management of serving equipment, recycling, and staff for serving the food (*Air Transport World* 1997).

5 Information provided by Robert Britton, Managing Director of Corporate Communications at American Airlines (May 1997).

6 Information provided by Robert Britton, Managing Director of Corporate Communications at American Airlines (May 1997).

7

DISTRIBUTION TECHNOLOGIES AND DESTINATION DEVELOPMENT

Myths and realities

Simon Milne and Kara Gill

Introduction

In recent years numerous researchers have focused on the ability of information technology (IT) to transform the tourist industry. Some commentators have stressed the intra-firm impacts of IT (Tourism Canada 1988; Haywood 1990; Bennett 1993; Pohlmann 1994) while others have presented more far-reaching analyses, arguing that IT is altering entire distribution networks and industry structures (Bennett and Radburn 1991; Mowlana and Smith 1992; Stipaniuk 1993; Go and Williams 1993; A. P. Williams 1993; Canadian Tourism Commission 1995). Indeed Auliana Poon (1993: 12–13) in her influential book *Tourism, Technology and Competitive Strategies* argues that IT changes the rules of the tourism 'game' – forcing all actors involved in the industry to adopt a new managerial and strategic 'best practice' (see also Poon 1988b, 1989).

Of the various tourism-related IT developments that have captured the imagination of researchers during the past two decades, computer reservation systems (CRS) have undoubtedly held centre stage (Collier 1989 and 1993; Tremblay 1990; Truitt *et al.* 1991). These powerful travel distribution technologies are seen by many commentators to offer significant opportunities for companies to improve customer service and increase the efficiency and flexibility of product delivery (Tapscott and Caston 1993; Jones 1993; Vlitos-Rowe 1992). Several researchers go as far as to argue that CRS is the dominant technology influencing the evolution of the industry (Go 1992a; Poon 1993: 13).

In this chapter we provide a review of recent attempts to understand the important role that CRSs can play in influencing the relationship between tourism and the destination development process. We begin with an overview of the technology itself and briefly trace its emergence as a dominant, global, travel distribution force during the past three decades. We stress the fact that there are several variants of CRS that must be studied, and that it is important to clarify the type of system and connection being used.

We then examine the work of Poon and other commentators who have tended to stress the positive outcomes associated with reservation-system development, including: improved corporate competitiveness; the creation of more 'flexible' and sustainable tourism products; and the re-skilling of tourism labour. We provide a critique of this literature, arguing that many of the positive impacts ascribed to CRS have been 'mythologized' and that we need to be cautious in attempting to study the links that exist between this powerful IT and destination development processes.

Our analysis begins with a review of the impact of CRS on firm competitiveness. Our findings show that this IT provides a series of important advantages for its users, and creates serious disadvantages for those that are unable to gain access. We then move on to explore the creation of more flexible tourist products/destinations and the degree to which CRS fosters alliance and network formation among firms. The important labour market impacts that stem from the evolution of this distribution technology are also presented. We then turn to the thorny issue of sustainability – in particular exploring the supposed ability of CRS to foster a greater degree of local ownership and control in the industry. Finally we present an overview of what the future holds for CRS in the age of the Internet.

In reviewing these issues we draw from a range of empirical work conducted by members and affiliates of the McGill Tourism Research Group (MTRG). Examples and cases are introduced from small island settings in the South Pacific, from the urban environments of Canada and Russia, and from Canada's eastern Arctic. Our primary objective is to show that certain trends and themes have a significance that transcends national boundaries and various destination types. In conclusion, we show the important role that locality and government can play in influencing the final destination outcomes associated with evolving travel distribution technologies.

Computer reservation systems – a review

The first CRSs were 'in-house' airline information and booking systems designed to handle flight schedules and rather limited changes to air prices: the inaugural system, American Airline's Sabre, began operation in the early 1960s (Collier 1989). At this time, the sale of seats on a flight usually involved a transaction between an airline reservation agent who would consult the internal CRS, and a remote travel agent. Travel agents searched through printed guides and tariff summaries for flight and fare information before telephoning the airline to determine availability and confirm other pertinent information (Scocozza 1989).

The labor-intensive nature of this process did not change until the late 1970s when CRSs began to make the leap from 'in-house' systems to remote, networked terminals (Chervenak 1992; Balfet 1993). This change was driven by several interrelated factors including the emergence of a range of new carriers in the wake of the 1978 US Airline Deregulation Act, and the resultant increase in the range of fares and schedules available to the public (Thomas 1990: 3; Truitt *et al.* 1991). Further justification for CRS development came from the realization that travel agencies, with their own dedicated terminals, could assume many of the information distribution and marketing tasks of an airline owner (Feldman 1992a). Not only did such 'farming out' of functions relieve carriers of the costs of providing the services directly, but it is well-documented that an agent utilizing a specific airline's terminal is more likely to choose that company's services – the so-called 'halo' effect (Feldman 1988; Sloane 1990; Ellig 1991).

By the early 1980s five major CRSs dominated the North American travel industry – Sabre (American Airlines), Apollo (United Airlines and affiliates), System One (Texas Air and Eastern airlines), Gemini (Air Canada and Canadian Airlines) and Worldspan (TWA, Northwest, Delta Airlines). A decade later over 95 per cent of airline seats booked in North America were handled through CRSs and 96 per cent of US agencies were using at least one automated booking system (*Travel Weekly* 1992). The 4 per cent of travel agencies that were not linked to a system were either 'technological dinosaurs' or were focused on niche markets that are not well-represented on CRS (Collier 1993).

Sabre continues to dominate the North American market-place – accounting for more than 40 per cent of all airline bookings through US travel agents in 1995 and generating $US225.9 million in profits from total sales of $1.53 billion (Canedy 1996). The system, which has recently been listed on the New York stock exchange, currently allows travellers to make reservations with more than 350 airlines, 55 car rental agencies and 190 hotel companies, and its terminals are found in more than 29,000 travel agencies in more than 70 countries (Canedy 1996).

The growing saturation of the North American market-place led US airline companies to globalize their reservation systems during the 1980s, with Sabre leading the penetration of the important European market. Until this time, European carriers had been relatively slow to invest in the development of the large-scale reservation systems that characterize the American industry. With the increasing drive toward the deregulation and privatization of European air travel, and the threat of domination by US CRSs, carriers such as British Airways, Air France, and Alitalia began in earnest to develop their own systems (Sloane 1990). The global expansion of CRS continued as American and European systems moved into the Asia-Pacific region. In response to this encroachment, a number of major airlines in the region initiated development of their own systems and alliances (Hepworth 1989; Truitt *et al.* 1991; Go 1995).

Airlines continue to seek out affiliations and alliances in order to increase their share of the world's travel retail outlets (Feldman 1992b; Knowles and Garland 1994; *The Economist* 1997). European, North American and Asian airlines have formed competing consortiums in an effort to imitate Sabre's success. For example, Amadeus/System One is a joint effort between Air France, Continental, Iberia and Lufthansa. As of 1994, three global distribution systems (GDS) were estimated to control 78 per cent of the world's share of hotel, car and airline reservations – with Galileo International capturing 28.3 per cent, followed by Sabre at 27.4 per cent, and Amadeus at 22.3 per cent (Tourism Canada 1994). Some sense of the potential for the future spread of these systems is provided by the fact that a very high percentage of travel agencies in parts of Europe and Asia remained unconnected to any CRS in the early 1990s (Table 7.1). There are, of course, also a range of non-airline based CRSs in existence. We now outline some of the major sectoral and regional networks that provide important data inputs into the GDS.

Table 7.1 Percentage of travel agencies with computerized reservation systems (CRS), 1992

South Korea	98	Singapore	56
USA	96	Spain	53
Australia	91	Taiwan	50
France	86	Germany	48
Italy	85	Malaysia	32
Japan	85	Philippines	32
Hong Kong	65	UK	23
Scandinavia	61	Greece	16

Sources: McGuffie (1994); Gill 1997.

Hotel CRS

The core of hotel CRS is the PMS (property management system), a storehouse of customer preferences, corporate accounts and contacts as well as data on room rates and availability (McGuffie 1994). PMS–CRS spread quite rapidly through the hotel industry in the 1980s – especially through expanding multinational chains (see McGuffie 1990) (Table 7.2). The costs inherent in the development of a hotel CRS are high, especially for independent operators. For this reason many independent enterprises do not develop their own systems, preferring instead to align themselves with a hotel consortium or hotel group representative (Archdale 1992). The Utell consortium, the world's largest, brings together over 6,500 subscribing hotels from 160 countries, and processes over 2.8 million reservations a year through its forty-four international offices (*Hotels* 1996).

Table 7.2 Percentage of hotels with computerized reservation systems (CRS), 1988

All hotels	Africa and Middle East	Latin America/ Caribbean	Asia and Australasia	North America	Europe
72.3	37.5	56.4	64.4	79.1	88.0

Source: McGuffie (1990)

Linkage of hotel CRSs with airline GDSs is vital. For the vast majority of properties this connection is made possible via 'switches' such as THISCO – a unified interface linking about 20 major hotel chain CRSs in the USA to airline GDSs (McGuffie 1994; Lindsay 1992b; Archdale 1992). In 1993 THISCO processed approximately 16.5 million hotel reservations for its linked hotels (McGuffie 1994; Tourism Canada 1994).

Regional reservation systems

Destination databases (DD), destination management systems (DMS), and regional/destination integrated computer information reservation management systems (R/DICIRMS) are just some of the terms used to refer to the regional reservation systems (RRS) through which potential travellers receive information focused on a specific destination or region (Buhalis 1993 and 1995; Canadian Tourism Commission 1995; Archdale 1992; Vlitos-Rowe 1992). The systems sometimes also have built in reservations capability. A DD is the simplest form of RRS – providing an information-only service to potential visitors (Tourism Canada 1994). More sophisticated systems, such as DICIRMS, offer the customer the ability to reserve in real time at the time of inquiry, and will also often boast a customer database that can be used by the system owners for the marketing of the destination.

The primary sponsors of RRS development are usually regional or national tourism offices (NTO) – often with the additional financial and/or technical support of local private-sector operators (Tremblay 1990; Archdale 1992). As part of the NTO's mandate, therefore, the RRS is expected to 'provide complete and up-to-date information on a particular destination . . . (and to) ensure that smaller establishments, as opposed to international hotel chains, and other land services are included' (Vlitos-Rowe 1992: 86; see also Sheldon 1993a and 1994).

While there are a number of important RRS success stories (see Vlitos-Rowe 1992), there are also a number of well-known failures including the recent demise of Reservations Quebec (Gill 1997). Buhalis (1993: 373) offers a comprehensive listing of the reasons for RRS failure, including: incomplete representation of tourism suppliers at the destination; inaccurate pricing and inventory; lack of finance for development and operation; limited tourist industry interest; weak training; inappropriate technology; and conflict between differing interests. The most formidable of these challenges is often limited financing (Sheldon 1993a; Archdale 1992). Because of these monetary constraints, NTO support for RRS development is increasingly restricted to the design and development phases, with the operational phase usually involving some form of privatization (Archdale 1992; Gill 1997).

Ticketing agencies

Ticketing agencies (TA) have existed for decades as intermediaries in the sale of tickets and passes to sporting and entertainment events (Ticket Master, previously Ticket Tron, began operating in 1967 in the US). The exploitation of powerful information technologies in the last decade has increased the TA's role in the marketing and distribution of a broader range of ticketed services: a traveller can now purchase passes to a museum, book a hotel room, and even buy an airline ticket (Gill 1997).

TAs enjoy a high degree of recognition from the public. Through aggressive promotion of their 1–800 phone numbers, these large concerns are easily identified and accepted by consumers. This accessibility is fostered through the placement of TA outlets in major malls and at some hotel front desks. TAs are also experimenting with ATM dispensers for ticket purchase. In some cases, international exposure is possible through connection of the TAs to GDSs (Gill 1997).

In addition to the sales service offered by TAs, advances in database management enable consumer profiles to be developed as each ticket is sold. This data is then available to travel managers in the form of market reports or in raw form (down-loaded to the travel product's own database). This information may, in turn, be used for niche marketing, and the revision of competitive strategies.

The implications of CRS development

Poon (1993: 12–13) points to a series of profound implications that IT, and the evolution of CRS in particular, hold for the tourist industry and those destinations that depend on it for their economic survival: 'IT facilitates the production of new, flexible and high-quality travel and tourism services that are cost-competitive with mass, standardized and rigidly packaged options.' She then goes on to state that: 'IT helps to engineer the transformation of travel and tourism from its mass, standardized and rigidly packaged nature into a more flexible, individual oriented, sustainable and diagonally integrated industry.' Poon is clearly emphasizing not just the cost-cutting dimension of CRS but also its ability both to cater to and stimulate more individualized forms of travel. Networks of small and large firms are efficiently and cost-effectively formed via CRS – allowing for the development of more flexible tourist packages and products.

The sustainability dimension stems from the perceived ability of CRS to increase levels of local participation in the tourist industry. As Poon (1993: 331) notes: 'true long-term sustainability will come from the development of an indigenous entrepreneurial class'. The degree to which local people are able to participate directly in tourism development is considered to be one of the most important factors determining the cultural deterioration and levels of local antagonism often ascribed to the industry (Macnaught 1982; Murphy 1994; Grekin and Milne 1996). It is also the case that increased local participation will improve levels of environmental awareness and stewardship, maximize local economic linkages and provide an important 'role model' for budding local entrepreneurs (Milne and Nowosielski 1997).

While there can be no doubt that IT has been vital in shaping the competitive stance of tourism enterprises and in influencing the distribution networks that link travellers to various destinations, we argue that it is necessary to look critically at some of the issues raised by Poon and other commentators who have tended to emphasize the corporate and destination benefits that flow from CRS adoption (Tapscott and Caston 1993; Jones 1993; Vlitos-Rowe 1992). In particular we want to look at the degree to which some of the 'myths' surrounding CRS development really match-up with the reality found in various corporate and destination settings.

CRS and corporate competitiveness

CRSs provide a good example of flexible service delivery systems that enable customers to make individual requests for specialized product combinations. There is little doubt that such systems provide the travelling public and the tourist industry with a very cost-effective way to develop a broad range of 'packaged' products. The price-sensitive leisure traveller can also benefit from the low-cost price searches available via a CRS (World Tourism Organization 1990; Kotler *et al.* 1993). In addition to market exposure, these systems also provide companies with detailed customer profile data, assisting them to direct better their sales to specific client bases.

In a recent study Gill (1997) attempts to study the impact of CRS on different dimensions of hotel and attraction competitiveness in Toronto and Montreal (Table 7.3). Not surprisingly, her dominant finding was that CRS is a key competitive tool for most of the enterprises surveyed. One of the hotel marketing managers interviewed by Gill summed things up when he stated: 'When you are in the system, you are in the game. If you're not in the system, you're not in the game.' Most enterprises simply have no choice but to belong to a CRS of some type.

This does not always mean, however, that companies are happy with the competitive performance of their CRS. While the majority of hotel and attraction managers surveyed by Gill felt that CRS had been effective in increasing their market profile and generating more business, they were

Table 7.3 CRS impacts on hotel/attraction competitiveness – Montreal/Toronto 1993–4

Competitive area	Average ranking (1–5)	Comments
Competitiveness (general)	1–2	Good
Bookings	1–2	Good
Market exposure	1–2	Good
Revenues	2–3	Acceptable
Labor productivity	4–5	Disappointing
Rate fixing flexibility	5	No impact
Alliance formation	5	No impact

Source: Gill (1997)

generally less impressed by the technology's ability to improve their room-rate and general price-fixing flexibility. At the same time, very few of those surveyed felt that CRS had improved their ability to compete through alliance formation – a point we will return to later in this chapter. It is also interesting to note that several companies found the revenue-related impacts of CRS to be less impressive than expected – a reflection, in most cases, of the relatively high costs of joining a system and paying the commissions that accompany each room-night or ticket sold.

Gill (1997) also found that perceptions of the competitive impact of CRS among hotels and attractions tended to vary according to firm size and ownership structure. The independent and smaller operations surveyed gave lower scores for the overall competitive performance of their respective systems than did managers of chain-related properties .

Flexibility and small/local firms

The ability of small, locally owned operations to gain access to CRS is a key element underlying Poon's assertion that IT is helping to create more sustainable and flexible tourism products. At the same time, however, several authors (including Poon) state that smaller, independent properties are those most likely to face problems in exploiting CRS technology (Buhalis 1995; Archdale 1992; Poon 1993; Pohlmann, 1994). According to Buhalis (1993: 370), while there are several reasons why small tourism firms usually lack access to IT (and CRS), most stem from a lack of effective organization, poorly trained staff and managerial weaknesses. While work conducted by the MTRG reflects these themes to some extent, the decision not to link up to a CRS can also be seen as a relatively rational one in many cases. The following listing ranks the major factors that preclude small and independent operators from CRS:

- high (often excessive) cost;
- not deemed necessary for success of their business;
- distrust of technology;
- desire to maintain control of inventory and sales;
- standardization of product listing.

Cost represents the overwhelming barrier to CRS adoption for small companies. The costs of booking commissions alone can exceed 20–30 per cent of the total room rate for small hotels, thus cutting into or eliminating any profit margin derived from the booking. One-off CRS fees can also be crippling for smaller operators (see Pohlmann 1994). At the same time, many small operators do not consider the CRS to be a necessary part of their business. Some fear that they will be overwhelmed by bookings generated by a system that lies out of their control, while other operators – especially in more remote areas such as the South Pacific and the Arctic – prefer to rely on traditional media (brochures, etc.) (Grekin 1994; Milne 1996).

Unfortunately, reliance on traditional forms of marketing is not always an effective competitive strategy for small operators trying to break into overseas markets. In their work on travel distribution systems and the development process in the Cook Islands, Burridge and Milne (1996) found that small, locally owned operations are poorly linked to overseas wholesaler packages that are marketed via brochures, while larger expatriate and foreign-owned operations are far better represented.

An even clearer picture of ownership structure and its impacts on CRS use emerges when chain affiliations are taken into account. In the urban Canadian work undertaken by Gill (1997) and Pohlmann (1994) it is clear that chain-affiliated or chain-owned properties have far greater access to CRS and GDS links than their independent counterparts. It is also true, of course, that some chains may themselves be laggards in adopting new technologies – placing whole sectors of certain tourism industries at a competitive disadvantage. A clear-cut example of this latter trend comes from the Russian hotel industry where joint-venture, chain-based operations have far greater CRS access than their Russian owned counterparts, many of whom were part of old Soviet-era travel chains such as Inturist (see Milne and McMillan 1997b). All of the joint-venture operations surveyed in Moscow and St Petersburg in recent MTRG research are featured on at least one system while only one Russian hotel was found to have such a link. Indeed, the manager of a large (1,800 bed), Russian-owned, two-star, St Petersburg hotel did not appear to know about the technology or its significance to the hotel industry (Milne and McMillan 1997b).

It must also be remembered that in some cases the introduction of CRS can be viewed as a threat to the high level of personal service that is a feature of smaller, often independent, up-market hotels, tour operations and attractions (see Haywood 1990). Distrust of the technology may also make the managers of some smaller operations reluctant to relinquish their inventories and reservations to the third-party sales force of a CRS company.

Finally, even as CRS permits greater flexibility and efficiency in travel-choice itineraries, the inability of some tourism suppliers to conform their product information to CRS standards continues to result in restricted selections for consumers. It is no coincidence, for example, that tour operators and their relatively complex products are some of the least represented components of the travel industry. Those tours that are listed tend to be all-inclusive and relatively standardized in nature, such as cruises and packaged hotel stays (Bennett and Radburn 1991). The rigidities of CRS display make it very difficult for niche tour and travel operations to market themselves effectively (see Milne and Grekin 1992).

In summary, while CRS is a vital competitive tool, it remains out of reach for many small operators. While the range of products offered on CRS continues to broaden, the organizational characteristics of the systems – especially their high cost, and requirements for relatively standardised information – tend to reinforce the supremacy of the larger, chain-oriented tourism firms. In some respects, the survival of the non-specialized, small-to-medium-sized, independent tourist enterprise may be more difficult than ever.

Alliance formation and networking

While Poon acknowledges that small industry players will face hurdles in linking up to CRSs, she counters by arguing that small-firm networking and public-sector intervention can assist in overcoming these hurdles (Poon 1993). She also argues that CRSs facilitate the diagonal integration of different components of the tourist industry, in effect fostering network formation and the creation of more flexible tourism products. There are two issues that need to be dealt with here: (a) does CRS tend to foster alliance formation?, and (b) do alliances and networks really make it easier for small firms to gain access to these important systems?

Alliance formation is without doubt one of the key avenues open to firms who wish to bolster their competitive position. Cooke and Wells (1989) and Phelps (1994) note that the major factors underlying the desire to form a network are: obtaining economies of scale in technology development; gaining access to distribution networks and foreign markets; and the need to solve common industry problems and 'exploit mutual opportunities' (see also Tapscott and Caston 1993). For large firms wielding considerable market share, alliances are another means of achieving power and the ability to coordinate the overall system to their advantage. For small and medium-sized firms, alliance activity helps overcome obstacles too great for independent firms to tackle alone.

Gill (1997) is one of the few researchers who have attempted to study the link between CRS access and alliance formation. Her study of hotels and attractions in Canada's two largest cities showed that the overwhelming majority of managers sampled did not believe that CRS membership had enhanced their ability to create new alliances and travel packages. They had not been approached by other tourism suppliers who were part of the same CRS, nor had they taken the initiative and approached other listed products. It is clear that local level alliances, which are often the most crucial to firm and destination competitiveness, are not built around the cornerstone of CRS. Even RRSs were shown by Gill to be of limited value in stimulating new alliance formation on the part of firms. Instead these alliances are likely to form because of less tangible factors such as trust, reciprocity and a conducive local business environment (often featuring public sector assistance) (see Hansen 1992; also Chapter 14 in this book).

This does not, of course, discount the ability of CRS to bring together new combinations of travel products. Tour operators and travel agents with access to CRS displays are able to develop a range of possible product combinations and package them cost effectively. The individual segments of these packages are, however, often unaware of the larger 'package' that they are linked to.

The ability of small firms to use networks and alliances in order to gain access to CRS must also be looked at critically. While there are examples of networks being formed (see Gill 1997) around similar enterprise types (for example bed and breakfasts) or specific destinations (often linked to government-aided RRS development), it is generally difficult to find examples of small companies who have joined together of their own accord and have successfully gained access to broader travel distribution systems. In the Moscow setting, for example, some alliance formation between Russian-owned hotels does exist and it was suggested by some managers interviewed that this may form a starting point for greater CRS access. However, fierce competition from foreign tour groups has led to rising levels of in-fighting and makes any long-term alliance formation appear rather unlikely (Milne and McMillan 1997b).

In simple terms, it appears that too much emphasis has been placed on the role of CRS as a force leading to, or enabling, alliance formation. While this powerful IT certainly assists in the creation of large-scale alliances, it appears to have only limited ability to overcome the locality or industry-specific factors (such as long-term development of trust, reciprocity and business culture) that can play such an important role in shaping successful alliance activity. There is also little hard evidence to point to the success of small-firm networking initiatives aimed at gaining access to these systems.

Labour impacts

The labour force impacts related to the introduction of IT by the tourist industry take on two key dimensions: the impact on worker numbers, and the impact on skill levels (Riley and Dodrill 1992; R. Wood 1992). Several commentators argue that technology offers the potential to relieve staff from the menial, repetitive tasks of running a tourist operation, so that they may address the needs of guests more directly and efficiently (GM Robinson Associates 1993). Indeed a key argument running throughout Poon's book is that IT, if managed properly, will tend to enhance skill and service levels – while creating relatively little employment dislocation (Poon 1993: 29).

The evidence from recent MTRG research reveals that the impact of CRS on labour use is influenced by the nature of the jobs themselves, and the nature of the CRS connection. There seems to be little doubt that front-line staff (those who deal directly with the public) are likely to find CRS developments providing them with more time to carry out their most important task – catering to client needs (Pohlmann 1994). It seems clear, however, that these systems will also lead to a slow down in employment growth (and perhaps staff reductions) in key back-office and 'behind-the-scenes' areas. In 1995 Canada's Via Rail completed a switch-over of its seating inventory and rates to Sabre, the first rail line in the world to achieve this complete GDS interface (Gill 1997). One result of the move was the laying-off of 50 per cent of their marketing staff. Traditionally, the company had used a mobile sales force that would travel throughout North America selling seats. With the new GDS reservation capabilities this sales staff was quickly made redundant. It should also be noted that TAs and relatively sophisticated RRSs also enable a range of tourism operators to effectively 'out-source' some of their marketing and sales functions. Even simple internal computer reservation systems can save some labour costs through the elimination of cumbersome (often manual) methods of sales and reservation tracking.

Determining the impact of CRS on labour efficiency and productivity is not an easy task. Our research shows that some managers equate increased efficiency with a reduction in the amount of manual entries made by staff, and therefore a reduction in human error. Others feel that the need to monitor incoming reservations through the CRS requires additional input from labour and hence incurs additional labour costs (Gill 1997). Here the type of CRS connection becomes an important consideration. A fax-confirmation connection, for example, is very labour intensive, necessitating additional steps in the monitoring of hotel rates and reservations. Conversely, the on-line, 'real-time' systems of large chain properties reduce the amount of menial data entry tasks required for reservations confirmation. Gill (1997) found that the former enterprises were the least satisfied with the impact of CRS on labour productivity while those with 'real-time' systems reported higher levels of satisfaction.

Destinations and sustainable development

To date there has been relatively little research into the impacts of the evolution of CRSs on tourist destination development (see A. Beaver 1992; Buhalis 1993 and 1995). Much of the work that has been conducted points to the fact that CRSs heighten the market-place profile of destinations which boast a number of well-linked (larger) travel companies, while limiting the opportunities for destinations characterized by small and medium, locally owned, establishments (A. Beaver 1992; Milne and Grekin 1992; Milne and Nowosielski 1997).

CRS-related biases in tourist distribution may have a number of significant impacts on the ability of destinations to achieve more sustainable forms of tourism development. If tourist flows are directed to large, overseas-controlled operations, there is every likelihood that local 'downstream' tourism businesses will be excluded (see S. Britton 1982; Milne 1992 and 1997). This, in turn, may well lead to a commensurate increase in economic leakages, especially in less developed settings. Larger operations also tend to be relatively less intensive in their use of local labour resources (see Milne 1987). A lack of local participation in the industry may also have undesirable cultural and environmental consequences. High levels of expatriate ownership and management may tend to alienate locals (Macnaught 1982), while at the same time local owners may be more likely to exhibit improved stewardship of the surrounding natural and cultural resource base (Milne and Nowosielski 1997).

The future of CRS?

One further question that must be asked is whether CRSs, and the travel agents who rely so heavily upon them, will remain as key links in the travel distribution system of the future? Indeed, some commentators argue that we are beginning to move away from the notion of a channel of travel distribution, and closer to the idea of a fully functioning tourism system based on communication networks (Go 1995).

The Internet is the largest and fastest growing public network in the world and represents an important new challenge to traditional CRSs. Not only is the Net registering 25 per cent annual growth in users, it also represents a lucrative market of potentially high-spending travellers. For example, US net subscribers have an average annual income of $60,000, 64 per cent have at least one college degree, and 50 per cent hold jobs classified as professional or managerial (Cyberatlas 1996; see also Hull 1996).

The ease and low cost with which users (public consumers, travel retailers) and vendors (tourism products, NTOs) can access the Internet has forced many CRS operators to establish World Wide Web sites. Sabre, for example, has introduced on-line EassySabre for consumers and Commercial Sabre for corporate use (Arnaut 1994). Use of these networks for actual reservations remains limited, however: EassySabre, for instance, generates only 1–2 per cent of Sabre revenues and Commercial Sabre 5 per cent (Gill 1997). Sabre has recently announced plans to debut an Internet version of its full CRS system, replacing its traditional hard-wired format. By moving its CRS customers to a web-based system, Sabre argues that agencies will save the time and expense of installing Sabre hardware, wiring and dedicated phone lines (Rosen 1997). It is unclear, however, whether Sabre and other existing CRSs will dominate the new area of on-line reservations. Airlines face considerable competition from new competitors such as American Express and Microsoft who announced an alliance in July 1996 to provide an on-line corporate reservations service (*The Economist* 1996).

The question of how these changes will affect the need for, or even the very existence of, travel agencies, has been the source of some debate (McGovern 1994). Some commentators have argued that the effect of new technology has been to reduce the customer/employee ratio and to displace jobs from individual travel retailers to centralized booking centres (Shaw and Williams 1994: 143). Others argue that the travel agent may be totally by-passed, and that we are simply witnessing the calm before a gathering storm of change (Walle 1996). Poon (1993: 193), on the other hand, in discussing the impact of new technologies on travel agents argues that they 'will not alter the manifest human content of neighbourhood travel agency services'.

Research undertaken by members of the MTRG suggests that Poon's rather rosy outlook may be somewhat misplaced, and that the picture becomes quite complex when one considers different travel markets. Pohlmann's (1994) recent survey of Montreal travel agents revealed that business travel agencies feel the most threatened by the growth of Internet-based technology, with 38 per cent claiming that corporate travel retailers will eventually disappear. Indeed, a total of over 60 per cent of business-oriented travel agents stated that they felt at least somewhat threatened by the technology. By way of comparison, only approximately 25 per cent of agencies catering to the general public felt threatened by on-line technologies (see also World Tourism Organization 1990: 14). While corporate travel offices and computer literate frequent travellers are currently the most likely groups to circumvent the travel agent, there can be little doubt that as Internet access becomes increasingly commonplace, and user interfaces are simplified, agents of all types will face considerable competitive pressure from this new way of accessing the travel distribution system (Canadian Tourism Commission 1995).

The increasing convergence of existing CRS technologies with the Internet makes it very difficult to determine technological boundaries (Canadian Tourism Commission 1995: 2). The growing interdependence and choice of reservation and information systems for travel suppliers and tourists make it highly unlikely that the dominance of the CRS systems of old will continue. It is equally likely, however, that new concentrations of control over travel distribution will emerge. The question is whether the leaders will stem from existing airline-based GDS, or from new players such as Microsoft (*The Economist* 1996).

Conclusions

In this chapter we have attempted to grapple with the links between CRS and processes of tourist industry and destination development. We have shown that in many respects CRSs do not represent a harbinger of new things to come and that, in fact, the past evolution of this travel distribution system has simply reinforced many of the patterns associated with what Poon would call 'old' tourism. The patterns of industry concentration and control that have emerged in the age of mega-CRS/GDS appear to be reinforcing the multinational corporate dominance found in the 'old' tourist industry, with a few large firms stamping their imprint on the direction and scope of future developments (S. Britton 1982). This supports Hepworth's (1991) broader claim that an electronic market-place based on proprietary networks will continue to support the domination of large globalized firms.

Larger, foreign-owned and chain-linked operations generally have greater access to CRS and as a result have a decided competitive advantage over their smaller, often locally owned, counterparts. These competitive biases will tend to facilitate against the achievement of higher levels of local ownership, and the creation of more sustainable forms of tourism development. There is, however, no doubt that CRS does allow the development of flexible packages in a cost-effective manner. In simple terms we appear to be witnessing the emergence of a competitive structure characterized both by increased concentration and by higher levels of fragmentation and segmentation (see Quinn and Gagnon 1986). We would argue that this form of tourism is neither entirely 'new' nor necessarily more sustainable than the industry that characterized the past two to three decades.

These broad processes of change are, however, not immune to attempts by the public sector to improve the market-place access of small and locally run businesses. Those destinations and regions that take a pro-active stance may well be able to temper some of the negative features of CRS growth that we have illuminated in this discussion. Areas of government/private sector input may include:

• the promotion of 'flexible standardization protocols' which can strike a balance between the generic language of computer communications and the variability of database information required to accurately describe tourism products;
• the development of technology information databases and training programs that can make small businesses aware of new ITs and how they may enhance the ability of small firms to improve their market profile;

- the highlighting of some of the potential benefits associated with network and alliance formation for small firms that wish to gain access to these systems.

It is also important to note that corporate and destination success in the global tourist industry will continue to reflect the unique local conditions that either foster or hinder profitable, appropriate, and sustainable tourism development processes. If there is one overarching theme that emerges from this chapter it is that tourism researchers must be wary of being too technologically deterministic and overly reductionist in their approaches to understanding the evolving relationship between the tourist industry, information technologies and the destination development process.

Acknowledgements

The research presented here was supported by grants from the Social Sciences and Humanities Research Council of Canada, the Quebec FCAR and the McGill Faculty of Graduate Studies. We would also like to thank Stephen Tufts of the McGill Tourism Research Group for his invaluable research assistance.

8

TOUR OPERATORS: THE GATEKEEPERS OF TOURISM

Dimitri Ioannides

Introduction

Over the last three decades, tour operators have emerged as key players in the international tourism arena. Together with the airline sector, these whole-salers of the travel industry are strategically placed as 'gatekeepers' exercising enormous influence over the geography of origin–destination tourist flows and, ultimately, the fortunes of individual destinations (S. Britton 1982; Burkart and Medlik 1981). Their competitive advantage derives from their pivotal position as coordinators responsible for packaging into single products the various elements serving the travel experience (i.e. flights, accommo-dation, tours, ground transfers), and selling these either through travel agents or directly to consumers at a single, discounted price (S. Britton 1991; Ioannides 1994; Middleton 1988; Urry 1990).

Yet despite the apparent crucial role played by tour operators in deter-mining the dynamics of international tourism, these actors have received surprisingly little attention in academe, especially compared to the airline and hotel sectors (see Daniels 1993; Debbage 1994; Go *et al.* 1990; Go and Pine 1995; United Nations Center for Transnational Corporations 1990; Wheatcroft 1994). Only a handful of tourism researchers have examined the structural characteristics of the tour operating industry, and the few existing comprehensive studies are quite dated (R. Britton 1978; Delaney-Smith 1987; Dunning and McQueen 1982b; Sheldon 1986; Touche Ross 1975). As Sheldon (1986) notes, there is a particular dearth of research relating to the North American tour-operating industry.

Within the context of this book, it is particularly important to note that although economic geographers have begun paying increasing interest towards service industries (Christopherson 1993; Coffey and Bailly 1992; Daniels 1993; Noyelle and Stanback 1984), they have entirely ignored the tour-operating sector. To a major extent, this silence reflects the overall tendency of economic geographers to overlook consumer services including most tourist-related activities (Ioannides 1995).

A fundamental reason behind the scarcity of academic research on tour operators, particularly in the United States, is surely the shortage of reliable data relating to the industry, a point made more than a decade ago by Sheldon (1986). It is almost impossible, for instance, to estimate the extent of the US tour-operating industry,[1] nor do we know which companies are the largest and to what degree the sector is characterized by industrial concentration. Though *Travel Weekly* publishes annual surveys of the travel agency sector, no comparable analyses exist for tour operators. Moreover, industrial surveys performed by companies such as Standard and Poor for almost every sector of the US economy entirely ignore the tour-operating industry. To make matters worse, the tour-operating industry is extremely secretive and company representatives appear reluctant to part with any information especially regarding volume of sales, trends in market research, or arrangements concerning commissions. Robert Whitley, the President of the United States Tour Operators Association (USTOA), attributes this caution to the extremely competitive nature of the tour operating business and the fear of industrial espionage by rival businesses.

The overriding purpose of this chapter is to examine tour operators as major players of the travel industry, given their ability 'to shift tourist flows from one destination to another or one supplier to another through the travel products they construct and promote' (S. Britton 1991: 457–58). These operators are more often than not in the driver's seat when it comes to negotiating prices with destination-based suppliers of tourist services. Importantly, major tour operators are able to easily substitute destinations in response to factors such as changing consumer tastes at the origin, or declining environmental conditions at the destination. Thus, the fortune of individual destinations can depend heavily on decisions and actions of a select few tour-operating companies in key origin countries.

The first part of this chapter overviews the composition of the tour operating industry. Attention is paid to the behavioral traits of tour wholesalers, particularly their decisions and strategies in dealing with other key players of the travel industry, namely the accommodation sector, travel agents, and governmental organizations in destination areas. Some key questions to be resolved include: What are the guiding forces influencing travel wholesalers when adding a new destination or travel product to their itineraries? Once they choose a new destination and/or product, how do tour operators interact and negotiate with hoteliers, transportation companies, or national tourist authorities? How easily can these operators substitute one destination with another?

The second major section focuses more narrowly on the specialty tour operators, particularly in the US. This sector has rapidly expanded in recent years, largely in response to growing consumer sophistication and the demand for increasingly flexible, custom-made, alternative package holidays as opposed to the standardized travel products traditionally associated with major wholesalers (see Chapter 6 in this book). Ironically, major tour operators have also responded to the demand for alternative tourism by offering flexible packages catering to a variety of market niches. These recent developments in the tour-operating industry have had important ramifications for the geography of tourist flows. Significantly, over a relatively short period of time the supply of specialized tour packages has transformed numerous remote areas throughout the world into popular destinations for alternative tourists.

Evolution of the package tour industry

The concept of packaging the various components of the travel experience into a single product is hardly new. Approximately 150 years ago, the Englishman Thomas Cook first began selling railroad package tours within Britain and later expanded his operations to destinations around Europe and eventually worldwide (Brandon 1991; Sheldon 1986; Shaw and Williams 1994). In the United States, around the same time, Wells Fargo (the forebear of American Express) entered the tour business by offering various travel-related services. During the early part of the twentieth century, package tours on steamships gained popularity. In 1926, for example, Thomas Cook sold three-week cruises from Cairo to Aswan and back for £80 (Brandon 1991). By the 1930s companies were selling inclusive air tours from New York to Florida, California, and the Caribbean (R. Britton 1978).

However, it was not until after the Second World War and, importantly, the significant developments in aviation technologies (namely the advent of the passenger jet aircraft), that the tour-operating business really expanded. During the late 1950s, airline companies realized that tour operators could help achieve higher passenger load factors by filling empty seats and providing these at significantly discounted fares in various combinations with other elements that make up a packaged inclusive tour. Concurrently, hotels began marketing their rooms through operators, 'since no sales staff or front-end money is required, and payment is made only for what business is produced' (R. Britton 1978: 133). Tour operators allow the various suppliers of tourism services to reduce their promotional expenditures (Sheldon 1986). Travel agents, the retailers of the tourist business, benefit from tour operators because they do not have to spend enormous amounts of time locating information on specific destinations or travel products. At the same time, consumers benefit because of convenience and reduced transaction costs.

Tour-operating companies survive through maintaining price competitiveness and providing good quality service. They strive to achieve large volumes of bookings in order to obtain significant discounts from their various suppliers. The major tour-operating companies can achieve a high number of sales because of operating efficiency and their financial ability to advertise. When constructing their travel packages, major tour operators commonly enter into agreements with brand-image hotel chains and airlines in order to promote and maintain a strong image of quality service provision. Similarly, they contract with experienced managers and guides to provide quality tours at their featured destinations. These factors indicate that the large tour operators can generate significant economies of scale. Alternatively smaller companies are able to remain in business by offering specialized services to a specific market segment (e.g. biking holidays, diving tours in coral reefs, adventure travel in polar regions, etc.) (Reimer 1990).

Tour operators play a vital role in determining tourist flows to particular destinations, especially in less developed countries. Obviously there are variations in the degree to which different countries depend on travel whole-salers. As Shaw and Williams (1994) point out, approximately only 30 per cent of British travelers to France participate in package tours. Alternatively, more than 80 per cent of travel from the UK to Greece is in the form of package holidays. Thus, the fortunes of numerous resorts particularly in the Mediterranean basin and also the Caribbean have come to depend quite heavily on the tour-operating business.

Due to their position as mass marketers of tour packages, tour operators function as key gatekeepers, often having a say as to whether or not a par-ticular destination rises in popularity (R. Britton 1978). In other words, the success of many countries as tourist destinations depends heavily on their ability to attract, and maintain over the long term, the attention of major operators. This is particularly the case with countries lacking a highly diversified tourist product and depending on a narrow range of substitutable resources (e.g. sand, sun, and sea). If, for whatever reason, operators lose interest in a particular destination, they can easily remove this from their itineraries and substitute it with another destination offering a similar range of resources (Goodall and Bergsma 1990).

But how do operators decide to market a new destination and/or product in the first place? Once they choose a new destination, how do the tour operators interact and negotiate with the key players of the travel industry (hoteliers, transportation companies, national tourist authorities)? The research in the next section draws heavily from secondary information sources and a survey conducted by this author in 1991–92. This survey sought to identify some of the key structural and behavioral characteristics of the tour-operating industry. Fifty questionnaires were mailed out to major European and US based tour operators yielding just fourteen responses. Subsequently, in-depth, open-ended personal interviews were conducted with representatives of four large tour-operating firms based in New York City.

The behavior and activities of tour operators

Choosing new products and/or destinations

Price is the single most important force guiding a tour operator's decisions (Goodall 1988; Reimer 1990). Most major wholesalers have traditionally concentrated on fixed markets and numerous proven destinations and/or travel products to minimize their risks. Thus, they rarely practice drastic product differentiation. Rather, any new product they decide to adopt is likely to be an adaptation of an existing theme.

In almost all cases, for a country or region to become part of a wholesaler's itinerary it must be well endowed with natural or other attractions (e.g. historical, cultural). A mass marketer, packaging holidays in the sun, for example, will be concerned with the availability of high quality sandy beaches and non-polluted waters. Alternatively, an operator specializing in ecotours might be more concerned about the availability of diverse fauna and flora or the existence of beautiful landscapes and superb views. Operators are aware that their principal target client group is usually far more worried about the cost and quality of the package as well as the type of holiday (often sun, sea, and sand) than the precise location. In essence, then, mass tour operators market holiday type rather than place. This holiday type can be found at a range of possible destinations. The inclusive tour, 'being primarily a holiday, is not place-specific in being tied to a single destination although a particular environment, such as a seaside resort or health spa, is a necessary component' (Goodall and Bergsma 1990: 173). It is thus essential for operators to minimize risks by offering a number of substitute destinations in their itineraries. Invariably, tour operators who cater to the sun, sand, and sea market will feature more than one destination in their brochures. They know that in the event of any problems at any one of their destinations (e.g. political strife, environmental pollution, high inflation) they can still rely on sales to other areas offering a similar array of attractions and facilities. In recent years, many European operators have been promoting relatively under-developed southern Turkey as the new sun, sand, and sea destination in the Mediterranean, recognizing that mass tourists were beginning to lose interest in the more established (and overcrowded) destinations (e.g. Cyprus, Greece, Spain).

It is also important to tour wholesalers that a destination be highly accessible in terms of time and costs. The availability of good airline connections to and from a destination is vital. Unless the wholesale firm owns an airline, it will depend on existing services by scheduled or chartered carriers. For instance, TBI Tours, a New York-based wholesaler specializing in the Far East, cooperates through joint marketing and block seat reservations with United Airlines, Northwest Airlines, and JAL, the carriers with the most extensive network in that region.

In cases where the choice for a new destination and/or product has to be made, a prime consideration for tour operators is an adequate supply of high quality hotels and ancillary facilities. While tour operators expect their destinations to possess a certain level of attractions, 'they are even more selective in their use of destination facilities, e.g. accommodation' (Goodall and Bergsma 1990: 172). A representative of a company specializing in tours to China and the Far East mentioned that before marketing a particular destination 'we might think twice if [that destination] does not have an adequate fleet of minibuses'. Another supplier who sells tours to North Americans visiting the Near and Middle East argued that it is important for a destination to have at least some restaurants that cater to western tastes and that tour guides must speak perfect English.

Many tour operators will only target countries with brand-name trans-national hotels because these offer a familiar product and, thus, a certain quality guarantee to most of their customers. In some destinations, operators may work with local hoteliers as long as they are sure that their product is mainstream and comparable to that offered by their larger transnational counterparts. Only on rare occasions, if a certain destination's primary attractions are exceptional, will some suppliers sacrifice an element of comfort in the tour because they feel confident it will sell to a special interest group. In response to numerous requests, one US-based firm catering to the higher end of the market added parts of Eastern Europe and Turkey to its program, even though the standard of accommodation and other ancillary facilities was perceived to be far below their clients' normal expectations.

Yet another major factor determining a wholesaler's selection of a new destination involves political and economic stability. Generally most operators do not offer tours to destinations with a volatile political atmosphere (e.g. most of Central America, Zaire, Haiti). Moreover, tour operators may rapidly cancel tours to countries or regions where war or civil strife have suddenly erupted. During the Gulf War in 1991, numerous US and European-based tour operators discontinued their sales to most destinations in the Near and Middle East because of the uncertainty of the situation (C.M. Hall 1994).

After targeting a new destination, the tour operator may seek to establish contact with various suppliers of tourism services, particularly the airlines and the lodging establishments. Some tour operators like Thomson in Britain control their own subsidiary airlines. Others negotiate directly with airline representatives. Often the larger wholesalers have a team of regional representatives whose primary responsibility is to negotiate with hoteliers a year in advance to receive volume discounts and allow sufficient time to print brochures (Sheldon 1986). These discounts vary from 10 to 50 per cent and, if they are high enough, then the operator can pass some of the savings on to the consumer.

In some instances, hoteliers approach operators directly or through trade shows. To reduce their risks, the tour operators will commonly make block reservations at a number of establishments rather than deal with a single hotel. The agreed-upon price per room normally depends on projected occupancy rates plus general economic conditions (Sheldon 1986). To avoid any financial loss due to cancellations of bookings, the operators normally negotiate a release date with their accommodation suppliers; otherwise they will most likely be subject to a cancellation penalty. American Express Vacations has a seven or fourteen day cut-off date option if it does not manage to sell all the rooms of a certain establishment. The date-release strategy places the smaller unaffiliated hoteliers at a considerable disadvantage as they might not be able to fill the rooms at such short notice, especially during the off-peak season; the backlash of a cancellation may lead to financial ruin.

Another form of agreement between operators and key suppliers is the block purchase. This agreement is not as common as the block reservation because it increases the risk for the operator. Under this form of agreement the tour operators are obliged to pay for the hotel rooms or airline seats reserved, and thus the onus is on them to increase their sales volume. Since the operator bears considerable risk, the unit price of the hotel room or airline seat tends to be lower (Sheldon 1986).

Surprisingly, given the size of the tourist industry, only a handful of tour operators undertake extensive market research to match various segments to new tourism products (Reimer 1990). While a number of tour operators argue that they carry out market research, in most cases this involves simple comparisons with the performance of their competitors and observations of their yearly bookings to identify past trends. All tour-operating companies send out their representatives to new areas that they may view as potential destinations. These representatives will report back on the quality of accommodation, ground transportation, and ancillary facilities. Often tour operators visit a particular destination following the invitation of a national tourist organization. However, tour operators generally believe that most national tourist offices play a minor role in influencing the operator's decision to sell a new destination or product.[2] This is probably because, as one source put it, national tourist offices (especially those in less-developed nations) are commonly perceived as disorganized, lacking sophisticated marketing skills, and having a weak understanding of their country's tourism product.

Some of the larger companies like Thomson and American Express engage in more comprehensive research including analyses of domestic economic factors and their impact on various target markets, analyses of destination developments, and regular surveys of bookings. More complicated techniques, such as product testing of a new tour package, are impossible to undertake because 'the product is always intangible and cannot be simulated for product testing in the way that a test batch of a new grocery product formula can be made up' (Riley 1983: 254). Nevertheless, Thomson commonly engages in test marketing by producing, distributing, and advertising a brochure and monitoring the response (Heape 1983).

The majority of tour wholesalers simply rely on instinct when deciding what destinations or products to sell (Reimer 1990). As one representative of a New York-based company put it, 'people will always travel . . . there is always a market of a certain size for a certain destination'. Another operator stated that it is unnecessary to carry out formal research because 'we know who we are selling to . . . there is a destination for everyone. If a destination is added to our itinerary we are sure that it will appeal to a certain segment.'

Marketing the inclusive tour package

Many people become interested in traveling to a certain destination after reading a travel article (e.g. in their Sunday paper), watching a movie or television travel program, or by word of mouth (R. Britton 1978). The imagery of northern Vietnam's waterways, portrayed a few years ago in the film *Indochine*, generated much interest in that part of the world. Media coverage can also have a negative impact on certain destinations. For instance, articles about terrorist attacks by Islamic fundamentalists against tour buses in Egypt have caused tour operators to cancel their itineraries to that destination.

Once they decide to go on holiday, many tourists choose to book a package tour. But how do they determine which package to buy? Because of the consumers' inability to inspect the elements making up the tour package prior to traveling (unless they are repeat visitors or know someone who has participated in the tour before), operators rely on various promotional strategies to market their products (Goodall and Bergsma 1990). The tour operators' most important marketing tool is the travel brochure which acts as a surrogate for the actual product on sale. In the case of mass tourism, where the travelers are more often than not concerned about holiday type than actual destination, the brochure plays an especially important role. 'How a destination is "marketed" in such brochures is crucial for the volume of holiday-makers attracted to that destination' (Goodall and Bergsma 1990: 174).

The travel brochure's value extends beyond its informational role. It is the packager's key ammunition for persuading potential customers, particularly those with limited travel experience, to purchase their product rather than that of their rivals. Since tour operators who market 'look alike' sand, sea, and sun destinations can rarely differentiate their products on the basis of price, they try to compete on the basis of quality. Therefore, it is necessary for them to convey an image of high quality facilities and services through their brochures.

The brochures are particularly important for inexperienced travelers who do not know what to expect from their holiday. Thus the operators fill the brochures with high quality, glossy pictures and carefully worded text, taking care to emphasize the positive aspects of their destinations. As Goodall and Bergsma (1990) state, an interesting feature of a mass packager's brochure is that it underplays any information pertaining to a specific destination's attractions. It is often hard to distinguish between destinations when browsing through a standard holiday brochure. Typically, the information supplied in a brochure relates to the type of accommodation, the length of stay for various packages, ground transportation, and travel arrangements to and from the destination. While tour operators clearly overplay the positive aspects of their products, the vast majority stop short of using clear-cut deceptive practices. This is because they rely heavily on repeat bookings and, thus, cannot afford to undermine the firm's reputation (Sheldon 1986).

Interaction of tour operators with receiving area governments/institutions

The previous sections have described how mass tour operators provide homogeneous package tours to more than one destination to minimize their risks. Thus, if a prospective customer is unsure about a locale, there is often a comparable, substitute package for another destination. The lack of loyalty which packagers display towards individual destinations means they have no quibbles about adding or dropping a country from their itineraries for a variety of reasons, ranging from reduced consumer demand to dissatisfaction with the quality of services. Thus, the tour operators are in a controlling position when negotiating terms with governments of receiving countries.

Either directly or through their various trade associations, tour operators frequently protest about rises in airport taxes, inefficient infrastructure, or environmental pollution in key resorts (R. Britton 1978). The United States Tour Operators' Association (USTOA) often makes requests to countries to lift visa restrictions and allow the free flow of tourism between nations. A few years ago the British trade association, TOSG (Tour Operators' Study Group) persuaded the Portuguese authorities to reverse the imposition of restrictions on charter flights to that country. Similarly, in 1989 the International Federation of Tour Operators (IFTO) launched a campaign to persuade authorities to adopt a tough stance against environmental and noise pollution in popular resorts. In one instance, a company selling tours to Zakynthos in the Ionian sea, started distributing leaflets to its customers warning them not to encroach on the habitat of the endangered loggerhead turtle.

While these actions on behalf of the tour operators and their lobby groups could be interpreted as attempts to embrace the banner of environmental consciousness, a more pragmatic interpretation is that they represent efforts to safeguard their profitability. If a destination becomes increasingly unpopular (e.g. because of overcrowding, overdevelopment, pollution), wholesalers will eventually substitute it with another less crowded destination possessing a superior environmental quality. Some tour operators do not even complain about problems at particular destinations; they just refocus their attentions towards a new locale. Asked if her company pressures national tourist offices or governmental agencies to curb environmental problems at a destination, one tour operator stated that 'it is a useless approach. We put other eggs [destinations] in our basket and pray a lot!'

Market concentration in the tour-operating industry

The tour-operating industry is extremely volatile, as displayed by the high number of firm entries and exits (Sinclair 1991). In the US, for example, between 1978 and 1985 1,008 new firms entered the business (Sheldon 1986). During the same period a substantial number of firms ceased operating altogether. As Sheldon (1986: 357) mentions ' . . . in 1978 there were 588 tour companies in the US; by 1982, that number had increased to 687, however, only 46 per cent (272 companies) were common to both years. . . . Only 34 per cent of the companies in existence in 1978 were also in existence in 1985.'

Similarly, in Britain, many of the companies established in the 1970s taking advantage of limited entry barriers and a surge in the demand for travel, no longer operate (Delaney-Smith 1987). In an environment of cut-throat competition, numerous firms which were unable to attract high volume bookings, could not afford to stay in business simply by slashing their prices. The only smaller tour operators that managed to survive were the ones targeting certain 'specialist' market niches.

Because profits in the tour operating industry are minimal, as low as £1–2 per £275 holiday (Urry 1990), operators strive to achieve large volume bookings, thus reducing the per-passenger costs. In addition to achieving economies of scale, large firms have a major incentive to stay in business because of substantial investments in information technologies and marketing. The largest companies are likely to belong to trade associations (like the USTOA) and will thus carry consumer protection insurance, meaning the travel agents are more likely to do business with them. The smaller operators can avoid direct competition by targeting specialized segments, though in recent years major firms have begun carrying more than one product and practicing more sophisticated forms of market segmentation by catering to a variety of specialized niches.

The large tour-operating firms' ability to generate considerable economies of scale and scope has led to a gradual polarization in the business through both horizontal, and to a lesser degree, vertical integration. Although in the United States and most western European countries there are now thousands of tour operating firms, only a handful of these control a major part of market share (e.g. Thomson in the United Kingdom, Neckermann und Reisen in Germany, and American Express Travel Services in the United States). As far back as 1975, when the last comprehensive study of the US tour-operating industry was undertaken (Touche Ross 1975), less than 3 per cent of all tour operating companies controlled 30 and 37 per cent of revenues and passengers, respectively. Similarly, in 1988 two companies (TUI and NUR) had gained control of the lion's share of the West German market. Collectively, these two operators were responsible for two-thirds of all annual tour sales in that country (Goodall 1988).

Between 1984 and 1988 the proportion of inclusive tour holidays sold by the top five operators in the United Kingdom rose from just over one half to almost 70 per cent (O'Brien 1990), and in 1989 had reached 77.5 per cent (Sinclair 1991). By 1993 the degree of industrial concentration had diminished somewhat, with the ten largest UK-based companies accounting for 70 per cent of total capacity (*The Economist* 1993b). Nevertheless, a high degree of industrial concentration still characterizes the industry. Just as in the airline and international hotel industry there have been numerous examples of horizontal integration between tour-operating businesses, one of the most notable being Thomson's takeover of Horizon in August 1988 (Urry 1990). Whereas Thomson's market share stood at 19 per cent in 1984, by 1988 this had increased to 32.5 per cent accounting for three million packages. By contrast, during the same year there were almost 700 operators in Britain who collectively accounted for only 30 per cent of total sales. Most of these smaller operators were specialist firms selling fewer than 10,000 holidays annually.

The move towards horizontal integration in the British tour-operating industry resulted after an intense competitive war among leading companies in the 1980s. In 1987, by slashing its prices by as much as 20 per cent over the previous year, Thomson Holidays managed to enhance its market share. Though Thomson's profits fell over the next two years, its subsidiary charter airline, Britannia, increased its profitability significantly, thus keeping the largest British tour operator afloat. Many other smaller companies that had also tried to cut their prices in order to stay in competition were less fortunate. A number of these smaller tour firms were unable to stay in business and were eventually bought out by their larger competitors. In the 1990s this volatility in the tour-operating industry became even more pronounced. In the early part of the decade, the International Leisure Group (ILG), which after Thomson Holidays had been Britain's second largest tour operator, was one of twenty-two operators to go out of business (*The Economist* 1993b).

Vertical integration has also become increasingly common in the British tour-operating industry with operators expanding into the airline, lodging, and travel agency sectors. For instance, Airtours owns an airline (Airtours International) and the Going Places travel agency. The degree of vertical integration in the UK tour-operating industry (particularly between the largest tour operators and travel agents) has created considerable concern for trade and consumer watchdog groups because it has led to a situation where consumers may be overpaying for their travel packages while facing restricted choices (Renshaw 1994; Marston 1996). The British Monopolies and Mergers

Commission has been called in to investigate the dominance of the four largest tour operators because of the potential abuse of market power. Consumer groups charge that the largest travel agencies are more than likely to promote the holiday packages of their own parent tour operators, even if independent wholesalers offer cheaper alternatives. 'Independent tour firms say the agencies charge them commission at least 50 per cent higher than normal and refuse to display their brochures if they do not pay' (Marston 1996 on the Internet).

Despite the lack of recent data relating to the size and composition of the United States tour-operating industry, wholesalers in the US also display market concentration through horizontal integration, though nowhere near the extent witnessed in Europe. Conversely, vertical integration, which is quite common in the European context, is extremely rare in the US tour-operating industry. Sunmakers, a Seattle based tour operator, recently acquired Maupintours, an upscale, escort tour operator based in Kansas (Del Roso 1996), adding to the nine other operators it has taken over since 1990. With this recent acquisition, Sunmakers expected to increase its annual profits from \$130 to \$150 million, ultimately serving approximately 125,000 people. Moreover, the company hopes to become a nationwide operator in the near future.

US-based tour companies rarely display the scale and magnitude of their European counterparts, a number of which sell more than a million packages a year. This perhaps reflects the fact that in the US the role of tour operators in the travel industry is not as great as in Europe (Poon 1993). According to the USTOA, any US-based company with more than 100,000 sales per year is considered large (but compare that to a number of British operators with sales of more than a million). Only a handful of American tour operators (e.g. AAT King's Australian Tours and GoGo Worldwide Vacations) have sales exceeding the 500,000 mark. All forty-four members of the USTOA[3] combined sold a total of 6 million packages in 1995, generating \$3.5 billion in sales.

Despite the differences, particularly in terms of size, between American and European tour wholesalers, the US tour-operating industry has become more concentrated, with a relatively small number of players controlling a significant share of the market. Nevertheless, in recent years there has also been a proliferation of small independent travel operators who have been able to survive in the business by catering to specific, specialized market niches. The rise of the independent travel specialists has been in response to the growing sophistication of the consumer and the increased dissatisfaction with conventional, inflexible tour products. These specialists have exploited the rising demand for ecotours, adventure travel, and other niche holidays. The final part of this chapter reports on the findings of a pilot survey of the specialist tour-operating sector undertaken by this author in 1995.[4] It summarizes some of the key traits of these travel specialists and seeks to identify how they have been affected by the recent revolution in information technologies including computer reservation systems (CRSs) and the Internet. Moreover, it explores the key challenges facing the specialist tour-operating industry in the next few years.

The US specialist tour-operating industry

General findings

The pilot survey of specialist tour operators took part during the summer and fall of 1995. The sampling frame chosen for this survey was the 1995 *Specialty Travel Index* (Volume 30) which lists a total of 600 operators. After selecting 250 firms at random, questionnaires were mailed out. In all, eighty-three firms (33.2 per cent) completed the survey.

The mean age of the responding tour companies was 16.4 years. In fact, 67 per cent had been created after 1980 (33 per cent had only been in existence for six years). By contrast, only 7 per cent of the companies had existed before 1950. This finding confirms the relatively recent emergence of the specialist tour-operating industry in response to the growing demand for alternative tourism.

Approximately 88 per cent of the tour companies which completed the survey were created as completely new ventures (many of them 'mom and pop' businesses), while the remaining 12 per cent were set up as branch operations of existing larger firms. Almost all the companies stated that they were independent, unaffiliated operators. Three companies were owned by airline companies, one had been set up as a subsidiary of a larger tour operator, one was owned by an auto manufacturer, and one by a travel book publisher.

All companies were asked to define the special interest category best describing their travel products. Roughly 23 per cent reported just one specialty (e.g. adventure tours, ecotours) while a further two-thirds mentioned that they specialized in two or more categories (Table 8.1). The majority of companies (73 per cent) described themselves as adventure-oriented operators and 31 per cent claimed they were ecotour specialists. Not surprisingly, almost 93 per cent of all companies indicated they were regional specialists meaning they concentrate in only one country or region. The majority of respondents specialized in locations within the United States with Alaska being the most popular region. While almost all these specialist tour operators sold only destination-specific packages (i.e. tours and accommodation), the majority (59 per cent) also included transportation to and from the destination. The principal argument for including the transportation component in the whole travel package is that the most significant cost savings that can be passed on to the consumer relate to transfers to and from the destination.

The vast majority of companies in the survey were small. Around 50 per cent had fewer than five employees while only three firms had more than 100

Table 8.1 Specialist tour operators by type of tour conducted

Type of tour conducted	Number of companies [a]
Adventure tours	59
Ecotours	27
Cultural tours	26
Hunting/fishing tours	13
Cruises	9
Biking holidays	8
Sports specialists	4

Source: Survey of specialist tour operators (83 respondents).
Note:
a Many companies reported more than one activity.

employees. Moreover, 57 per cent of all respondents had only one office and only three companies had more than three. Even more interesting was the fact that, unlike the major players in the industry which control a substantial share of the market, most of the companies in this survey sold a fairly limited number of tours. Almost 50 per cent of the tour operators reported fewer than 1,000 clients per year, while a further 36 per cent sold between 1,000 and 5,000 packages (Figure 8.1).

The reported sales volume of these companies ranged from $20,000 to $150 million per year. Interestingly, despite their small average number of clients, 30 per cent of the companies in the survey generated annual sales exceeding $1 million. This confirms the relatively high expense of specialized holidays (over $3,000 per person) (Weiler and Richins 1995). The high cost of specialized holiday packages is not surprising considering such products are commonly marketed toward the upper end of the market. Most companies indicated their products are geared towards older age groups (the 40+ category), college-educated persons with annual incomes above $40,000. Some companies even stated that they cater to individuals who earn in excess of $100,000 per annum. The results of this study are consistent with earlier research that has shown that consumers of special interest tours tend to be mid-to-upper income, college-educated professionals, the majority of whom are middle-aged or older (Eagles and Cascagnette 1995; Weiler and Richins 1995).

Operators were asked to assess their operation's performance for the past fifteen years or since their inception, based on changes in their clientele size and/or receipts over time. A number of companies failed to respond to this question reflecting perhaps the secretive nature of the business. Of the tour operators who did respond, most reported an increase in clientele size (or sales volumes) over the years. This may demonstrate the recent increase in demand for special interest holidays. In view of the fact that each year there are more

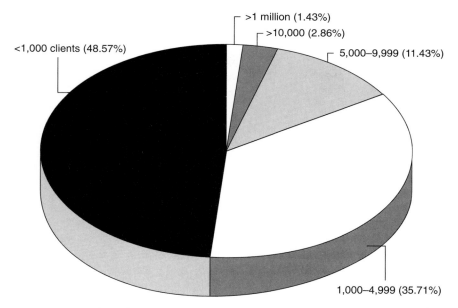

Figure 8.1 Number of tour firms by clientele size
Source: Specialist tour-operator survey

specialist operators, it shows that this niche of the travel industry may not yet have reached saturation.

One key issue that this survey tried to establish is whether, unlike major wholesalers, specialist tour operators bypass the traditional sales avenue through travel agents and sell directly to consumers. The findings indicated that approximately 11 per cent of the tour companies sold their products directly to clients while only 3.6 per cent (three companies) marketed their packages through travel agents only. The overwhelming majority of companies (85 per cent) sold their products both directly and via travel agents. A number of operators who use both marketing approaches indicated that the overwhelming proportion of their sales results from direct contact with their customers.

Direct marketing is attractive to specialist tour operators because they can reduce their costs considerably by not paying commissions to travel agents. The majority of consumers of specialist tours are sophisticated travelers who are willing to avoid travel agents, most of whom only sell conventional tour packages. These savvy travelers make their own travel arrangements or go out of their way to find travel specialists (Walle 1996). The clients of specialist tour operators fit the profile of Plog's (1991) allocentric travelers and most have already had considerable travel experience. Unlike their mid-centric or psychocentric counterparts who seek the familiarity offered by an all-inclusive travel product, and are often more concerned about product type rather than actual destination, allocentrics seek experiences in specific out-of-the-ordinary destinations. If such travelers want to go on a hiking holiday in the Himalayan region of India, rather than contact a traditional travel agent, they are often better served by a tour specialist who has built a reputation for designing custom holidays in that part of the world. Most specialist tour operators are intimately familiar with the destinations they market and are strongly positioned to answer any questions concerning local arrangements. Importantly, unlike their mainstream, large-scale, tour-operating counter-parts, travel specialists display considerable loyalty towards the destinations they market. After all, their entire business depends on specialized tours to a narrow range of destinations and, while they have the ability to construct custom holidays for individual travelers, they cannot easily substitute their destinations.

Reliance on technology

Direct marketing techniques appear to have benefitted from the recent revolution in information technologies, including advances in computer reservations systems (CRSs) and the proliferation of the Internet. As Milne and Nowosielski (1997: 16) argue, 'rapid improvements in information technologies are changing the relationship between consumers and travel suppliers and may offer new opportunities for small, niche-oriented products to reach potential markets'.

The specialist tour operator survey revealed that only sixteen companies (19 per cent) were linked to a CRS. Ten of these companies were linked solely to an airline CRS like System One. An additional four companies were linked to more than one CRS (e.g. airline plus hotel and/or car rental). The low percentage of specialist operators who have access to CRSs is not surprising. As Milne and Gill (see Chapter 7 in this book) argue, tour operators in general are the least-represented segment of the travel industry on CRSs. The survey finding appears to confirm Milne and Gill's belief that a number of factors, among them high costs, distrust of technology, and desire to maintain control of inventory and sales, inhibit small, independent travel specialists from using CRSs. Moreover, as Milne and Nowosielski (1997) maintain, the CRS display protocol is fairly rigid, meaning that it is easier to represent fairly standardized tours. By contrast, the flexible products of specialized tour operators do not easily lend themselves to CRS use.

While CRS technologies do not feature prominently in the specialist tour-operating industry, it appears that other forms of computerization are becoming very important. At the time when the survey was carried out, thirty-two firms (38.6 per cent) had established, or were in the process of creating, their own web page on the World Wide Web (WWW). Moreover, since the *Specialty Travel Index* has now created its own web page, at least basic information about all these companies (e.g. contact person, address, telephone and fax numbers) now appears on the Internet. This finding indicates that because of the Internet, a number of small-scale companies can establish direct contact with their clients, thus bypassing travel agents.

> Web sites offer the possibility to break out of some of the rigidities associated with CRSs while the increased use of home computers and growing accessibility of computer networks such as CompuServe may play a role in the demise of traditional non-specialized travel agencies.
>
> (Milne and Nowosielski 1997: 17)

A cursory examination of the web sites of some of the specialist tour operators in the survey reveals that the majority use the WWW for providing information and promotional purposes, very much as a 'virtual brochure'. The main advantage is that this 'brochure' can be accessed directly without paying a visit to the travel agency.

Challenges facing the tour-operating industry

The last part of the specialist tour operator survey gauges the respondents' opinions regarding the travel industry's future. Moreover, it examines the implications this perceived future may hold for different tourist destination areas.

An overwhelming majority of the operators (83 per cent) believe that conventional mass tourism will eventually be eclipsed by a new variety of tourism based on heightened levels of consumer sophistication and enhanced market segmentation. There is a general opinion that tourists have become much more sophisticated in recent years compared to a decade ago and that, increasingly, the actual travel experience more than the financial cost is becoming the principal determinant of holiday destination choice. Nevertheless, 53 per cent of all respondents do not think conventional mass tourism will reach saturation any time in the near future and there is still much room for expansion in this form of travel. While certain mass tourist destinations (e.g. traditional Caribbean and Mediterranean resorts) may continue to lose their popularity, others constantly emerge to take their place (e.g. islands in the Indian and Pacific Oceans). Similarly, though an increasing number of western Europeans and Americans may display shifting preferences toward individualized travel forms, the rapidly expanding Far East and South-east Asian markets still adhere very much to the mid-centric and psychocentric traveler profiles.

Asked about their thoughts concerning major challenges to the specialist tour industry, a number of respondents wrote about the principal threats to their companies' existence. A major concern is the increasing consolidation of the tour-operating industry with the large wholesalers controlling an enormous proportion of the market. A number of specialists expressed the worry that most principals have begun practicing product segmentation, creating flexible, specialized travel packages, catering to a multitude of markets. The specialist operators regard this trend as a threat to their own business. While they have to compete with numerous other specialists offering variations of similar products to a similar range of destinations (e.g. Costa Rica or Nepal), they now also have to worry about the large conglomerates who have entered the specialized niche game. This puts the specialists in a situation where, as one respondent mentioned, they have to 'constantly reinvent the wheel by searching for new products in the same destinations'.

Ironically, this situation has made many tour specialists staunch proponents of environmental protection in the destinations they serve. Many fear that as more and more large tour operators begin to sell ecotours or other specialized packages, many fragile areas will be opened up to mass ecotourism and eventually suffer irreversible damage. The emphasis placed on environmental protection by travel industry specialists in their key destinations is an attempt to protect their products, since they do not enjoy the luxury of their large-scale counterparts who have the ability to substitute destinations and/or travel products.

A final list of tour operators' concerns includes the excessive commission demands by travel agents, political and/or economic problems in some of their key destinations (particularly for those that serve destinations in the Third World), and increasing concern about government regulations on the tour-operating industry. For instance, similar to the Travel Directive which unifies the method of consumer protection around the European Union (Hodierne and Botterill 1993), most states in the US now require tour operators to place a bond in the form of an escrow meaning that tour operators cannot access the money paid for selling their travel packages until after trip completion. This means that tour operators stand to lose a major source of income, since they have traditionally relied heavily on the interest accrued from pre-payments for their travel products.

Conclusions: implications for destination areas

The tour operators' strategic positioning between the various suppliers comprising the travel industry, and between these suppliers and consumers, has traditionally given them an enormous competitive advantage. Through their ability to combine the various elements of an amorphous industry into relatively affordable, standardized, travel products, tour operators have emerged as key manipulators of tourist origin–destination flows. Major tour-operating firms have traditionally displayed little loyalty to any specific destination, preferring instead to employ a multi-locational strategy. To reduce their risks, rather than promoting any specific locale, these companies market holiday type which can be found at numerous destinations. Thus, tour operators can easily substitute one destination with another to maintain their profits.

Unfortunately, the tour operators' strength places a number of destinations, particularly those with a narrow range of tourism products, at a considerable disadvantage. Specifically, governments and tourism suppliers at these destinations (e.g. lodging establishments) often find themselves in a very vulnerable position at the bargaining table with the major transnational tour operators. For example, authorities in an island microstate may think twice before imposing restrictions on the number of arrivals on low-cost charter flights, anticipating that if this measure angers representatives of the tour-operating industry, the latter may easily remove the destinations from their itineraries, in turn leading to low occupancy rates and ultimately economic decline.

Moreover, since the smaller, locally owned hotels in numerous destinations, particularly in less-developed countries, are seldom linked to major CRSs and have limited abilities to market themselves abroad, they are often by-passed by mainstream tour operators and travel agents who prefer to cooperate with recognized transnational hotel chains. In those cases where local small-scale hoteliers cooperate with foreign tour operators, the latter are well placed to extract favorable room rates and high commissions.

In view of these factors, what can the destinations which rely over-whelmingly on tour operators do to secure the future success of their tourist industries? Does the substantial power held by the major, oligopolistic, tour wholesalers negate such a possibility? To an extent, answers to these questions depend on the future of the tour-operating industry itself, particularly in light of the growing dissatisfaction by many travelers for standardized, mass-oriented package tours and the emerging demand for more flexible travel arrangements. Whereas it is highly unlikely that mass tourism will be eclipsed any time soon, the growing demand for alternative tourism is having an impact on the tour-operating industry, as evidenced by the proliferation of specialty tour operators in recent years. Unlike their large scale counterparts, these travel specialists are not in a position to easily substitute destinations given that their business often depends entirely on one region. Thus, these players do not possess the traditional, strong bargaining power of the major tour firms. In the meantime, the ability of large-scale tour operators easily to switch destinations may also be waning, partly because there are only so many substitutes to choose from and partly because many travelers are increasingly savvy about specific destinations. Thus, tour wholesalers and travel agents may eventually find it increasingly difficult to sell products based solely on holiday type and financial costs as opposed to actual destination.

Despite these speculations, it is hard to say with any degree of certainty what the future holds for the tour-operating industry and, in turn, what this may mean for specific destinations. On a cautiously optimistic note, if the traditional bargaining advantage of tour wholesalers over numerous destinations begins to decline, tourist industry players in these destinations could strengthen their ability to dictate tourism development on their own terms. Most likely, the opportunity arises for a certain degree of collaboration between tour wholesalers and tourist industry players at individual destinations, given the need of the former to protect the quality of their tourism products and the need of the latter to maintain their competitiveness.

Acknowledgments

I would like to thank Southwest Missouri State University for a summer fellowship in 1995 which enabled me to undertake a major part of this study. Thanks also to Dr Milton Rafferty for providing the funds for the specialist tour operator survey.

Notes

1 By contrast, the British tour-operating industry appears to be monitored on a fairly regular basis and it is fairly easy to acquire statistics relating to the industry (see for example *The Economist* 1993a and 1993b).
2 Some suppliers argue that certain national tourist organizations are more active than others. Israel, for example, has a very active tourist board and often subsidizes the advertising campaigns of a few major suppliers. The governments of Singapore, Hong Kong, and Japan also spend vast sums to promote their tourist products. Alternatively, in a country like Turkey the national tourist board has a modest budget and is not in a position to finance extensive promotional campaigns.
3 To become an active member of the USTOA (the elite organization of tour operators in the United States), a tour operator has to have eighteen references, be in business for at least three years under the same ownership and management, meet certain minimum standards in terms of passengers and/or dollar tour volume, carry $1 million of professional travel agent/tour operator liability insurance with worldwide coverage, and post $1 million in the form of bond, letter of credit, or certificate of deposit to be used for reimbursing customers in case of member bankruptcy or insolvency. Travel agents overwhelmingly prefer to interact with USTOA members when choosing packages (Sheldon 1986).
4 This pilot survey is part of an ongoing study of tour operators began in the early 1990s, initially as part of my doctoral research.

9

THE AIRLINE INDUSTRY AND TOURISM

Stephen Wheatcroft

Introduction

The tourism industry in many countries of the world has been profoundly shaped by the development of air services. There have been both positive and negative influences. On the positive side, the advances in aircraft technology, improvements in communications and information technology, and more sophisticated management and marketing techniques have improved the quality of air travel and reduced the price of air tickets so that the volume of traffic, particularly on longer routes, has doubled in each of the past three decades (Organization for Economic Cooperation and Development 1997). On the negative side, international air transport has historically been a highly regulated industry with controls on routes, capacity and tariffs. Despite recent moves to liberalization, air transport is still subject to regulatory constraints in many parts of the world and certain aspects of international regulation, particularly those designed to protect state-owned airlines, continue to have adverse impacts on the development of tourism (World Travel and Tourism Council 1997a). The relationship between the airline industry and tourism is therefore a highly complex subject involving an intriguing mixture of technological factors, market pressures and regulatory policies which this chapter attempts to unravel. But before examining these issues it is necessary to take a look at the broad framework of the total travel and tourism industry within which these developments have taken place.

Travel and tourism – the big picture

Air transport is part of a broader travel and tourism sector which is now widely recognized to be the world's largest industry. This travel and tourism industry includes all economic activities which are dependent on the expenditures of travelers before, during and after their trips. This broadly based definition of the industry embraces activities in the following seven fields:

159

1   Transportation.
2   Accommodation.
3   Catering.
4   Retailing.
5   Recreational activities.
6   Cultural activities.
7   Travel-related services.

The total economic impact of travel and tourism can only be assessed by measuring current and capital expenditures in each of these fields including those by consumers, businesses and governments. The World Travel and Tourism Council (WTTC) has worked for several years with the WEFA Group (formerly Wharton Econometric Forecasting Associates) to develop a methodology by which each of these components of the industry can be measured. This involves a detailed review of national accounts to identify the percentage of each sector of the economy which depends upon the demands created by travel and tourism. The 1996–7 *WTTC Travel and Tourism Report* (WTTC 1996a) presents world-wide and regional estimates of the total economic contribution of the industry measured in this way. In summary the report concludes that travel and tourism in 1996 had a gross output of $3.6 trillion, which was 10.7 per cent of world GDP, and that it generated 255 million jobs which was 11.1 per cent of total world employment. It further estimated that travel and tourism was responsible for 11.9 per cent of all capital investments and that it contributed 10.4 per cent of all taxes paid to governments.

These introductory comments about the size of the industry establish a broad perspective about the economic importance of travel and tourism and explain why governments throughout the world are increasingly recognizing that they should foster the development of these activities. The relationship between air transport and tourism, and particularly the development of regulatory policies for the two sectors, has to be viewed within this larger economic picture of the whole industry.

Air travel and tourism

Within the total world-wide travel and tourism industry, the air transport component is relatively small. This is because transport by private car comprises such a large part of total travel in two major regions, North America and Europe. Over 80 per cent of domestic travel within the United States, and a similarly large share of travel between the United States and Canada, moves by private car. And in Europe, where travel between contiguous countries is increasingly difficult to distinguish from domestic travel, the private car also dominates the travel market with almost 70 per cent of the total (Wheatcroft 1994). The airline share of the market increases on longer routes for the obvious reason that time-saving is greater.

For this same reason, air transport comes into its own and eclipses other modes of transport on intercontinental routes. In the long-haul travel markets, airlines have now won very large shares of total travel. On North Atlantic routes, trans-Pacific routes and routes between Europe and the Far East, travel by sea nowadays has virtually been eliminated in the past few decades (Figure 9.1). Travel by sea is almost entirely confined to cruising (World Tourism Organization 1996).

The importance of air transport for tourism is illustrated in Table 9.1 which lists twenty-six countries in which 70 per cent or more of international tourist arrivals came by air in 1994. In fifteen of these countries virtually

Table 9.1 International tourist arrivals by air, 1994[a]

Country	Total arrivals (000)	Per cent by air (%)
Dominican Republic	1,717	100.0
Jamaica	977	100.0
Maldives	280	100.0
Martinique	419	100.0
Puerto Rico	3,042	100.0
Saint Lucia	219	100.0
Saint Maarten	586	100.0
Trinidad	266	100.0
Australia	3,362	99.7
Sri Lanka	2,127	99.6
Taiwan	2,127	99.6
New Zealand	1,323	99.5
Philippines	1,574	98.5
Mauritius	401	98.3
Seychelles	110	98.2
Kenya	863	95.0
Malta	1,176	92.7
Vietnam	1,018	92.4
Cyprus	2,069	90.2
Nepal	327	88.4
Singapore	6,268	84.0
Greece	10,713	82.4
Thailand	6,166	82.3
China	5,182	81.5
Korea	3,580	72.5
United Kingdom	21,034	70.0

Source: World Tourism Organization (1996)
Note:
a Countries with over 70 per cent arrivals by air.

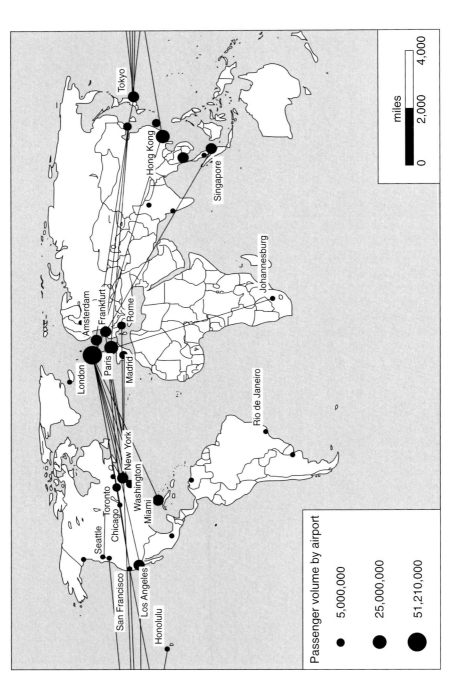

Figure 9.1 International scheduled airflows of more than 300,000 a year between pairs of the world's largest airports, 1991
Source: Debbage, 1994

all visitors arrived by air (World Tourism Organization 1996). The most interesting feature of this list of countries is those destinations whose development was dependent to a major extent on air services. This applies to most of the holiday islands of the Caribbean which rely almost entirely on air services for their visitor traffic. Of particular interest is the Dominican Republic where a remarkable growth in tourism has been promoted by charter air services from North America and Europe. In the European region, it was the expansion of inclusive tour air services in the 1970s and 1980s which was responsible for the large growth of the tourism industries of Greece and Cyprus (also of Spain and Turkey, even though they are not listed in Table 9.1.). In other parts of the world, new tourist resorts like the Seychelles, Mauritius and the Maldives have been entirely reliant on air services for their development. New developments to be monitored in the near future are the growth of tourist traffic by air to China and Vietnam.

It is clear that there is a powerful synergy between the development of international air transport and international tourism. The closeness of this relationship is illustrated in Figure 9.2 which relates international airline passenger revenues and international tourism receipts to world GDP over the

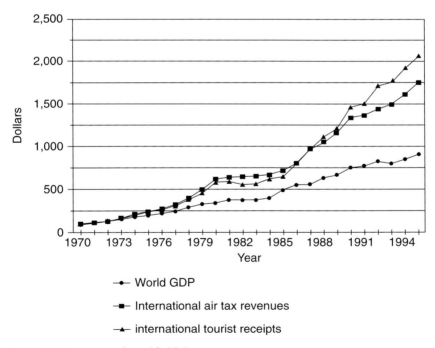

Figure 9.2 Tourism and world GDP
Sources: The WEFA Group, World Tourism Organization, and International Civil Aviation Organization

past twenty-five years. Figure 9.2 shows that, on a world-wide basis, spending on both international air fares and international receipts from tourism has increased at about twice the rate of world GDP growth.

It has become increasingly apparent in recent years that government aviation policies, which are designed to protect the interests of national airlines and which inhibit the growth of tourism, are likely to have adverse economic consequences for countries that pursue such policies. This is the central theme of *Aviation and Tourism Policies: Balancing the Benefits* commissioned by the World Tourism Organization (Wheatcroft 1994). The objective of the book was to produce a methodology by which all countries, rich and poor, can measure the advantages and disadvantages of alternative aviation and tourism policies and find the best balance. Analyses of this kind are vitally important because, despite strong pressures towards the liberalization of international aviation regulation, protectionist aviation policies persist in many parts of the world.

The liberalization of aviation regulation

In the past, almost all countries pursued aviation policies based on bilateral agreements designed to protect what were thought to be vital national interests (Table 9.2). These included the protection of employment, the promotion of hi-tech activities, defense considerations and national prestige. A national airline was a symbol of independence. The whole structure of international aviation regulation under the 1944 Chicago Convention was based on concepts of national sovereignty over airspace and the 'ownership' of rights to carry traffic.

This is now changing. The widespread adoption by many governments of 'economic disengagement' policies is leading to the reduction of regulation in many industries, including air transport. Domestic air transport operations have been deregulated in many countries and governments are increasingly withdrawing political controls from airlines. Privatization and deregulation are leading to a domestic industry structure in which airlines act as commercial undertakings in a competitive market (Wheatcroft 1990).

In the international field, the bilateral regulatory system which rigorously controlled air operations in the past is now in the process of radical change. Perhaps the most remarkable change is the creation of a multinational free-trade aviation area in the European Union. The so-called 'third package' of aviation liberalization, which became fully effective on 1 April 1997, creates a regime in which all European airlines are free to operate on any routes, including domestic routes, within the EU with no restrictions on pricing and capacity. It also provides for:

- Harmonization of national licensing rules.
- Community ownership rules for EU airlines.

Table 9.2 The language of aviation regulation

Bilateral agreements	Air service agreements between two countries regulating the routes to be operated, the airlines to be designated to operate those routes, the capacity of service, the fares and rates and other issues.
Multilateral agreements	The adoption of regulatory provisions by a group of countries.
Freedoms of the air	Rights defined by the Chicago Convention for inclusion in bilateral agreements.
First:	Overflying another country without landing
Second:	Landing for technical reasons.
Third:	Setting down passengers, freight or mail.
Fourth:	Picking up passengers, freight or mail.
Fifth:	Setting down or picking up traffic in a third country (e.g. a US airline carrying traffic between UK and Italy).
Sixth freedom	The combination of third and fourth freedom rights to gain access to other markets (e.g. a UK airline carrying traffic from New York to Italy via London).
Cabotage	The carriage of domestic traffic within another country.
Capacity control	Any form of agreement to limit the frequency of service or number of seats which may be operated by the airlines of either party.
Tariff control	Any form of agreement to determine the fares and rates charged by the airlines of either party
Double disapproval	A liberalisation measure under which a tariff proposal becomes effective unless disapproved by both countries.
US deregulation	The policies initiated by the US Airline Deregulation Act of 1978 which provided for unrestricted route entry and the abolition of capacity and tariff controls.
EU third package	The agreement reached in the EU by which all entry (including cabotage) capacity and tariff control would be finally abolished for air services within the EU in 1997.
Code sharing	The use of the same flight designator by two or more airlines for connecting flights.

- Full cabotage rights on EU domestic routes.
- Unrestricted fifth freedom rights.
- Free pricing subject to consumer safeguards.
- Competition rules to cover domestic services.

European airlines are free to set up operations in any other EU country, and the whole concept of 'substantial ownership and effective control' has disappeared. Multinational ownership will increasingly become the norm.

Another radical change is the acceptance by many European countries of an 'open skies' agreement with the United States. These agreements eliminate controls over pricing and capacity and give the European airlines greater freedom to serve more cities within the United States. Although these agreements are currently opposed by the European Commission, they may well become the stepping stones for a much wider, multilateral agreement between the North American Free Trade Area and the European Economic Area.

In other parts of the world, there are also moves towards regional liberalization. Australia and New Zealand have concluded a 'Single Aviation Market Agreement' which links the two countries across the Tasman Sea. Under this agreement the airlines in each country gain unrestricted rights to fly anywhere in the other country and on trans-Tasman services. In South America, aviation liberalization is one of the objectives of the Mercosur agreement (Argentine, Brazil, Paraguay and Uruguay) and of the Andean Pact (Bolivia, Colombia, Ecuador, Peru and Venezuela). Finally, in South-east Asia, the six members of ASEAN (Association of South-east Asian Nations) – Brunei, Indonesia, Malaysia, the Philippines, Singapore and Thailand – are developing a Free Trade Area which looks forward to a single aviation market in that area.

Throughout the world, therefore, there are multinational developments which are leading towards liberalized aviation markets within those regions (World Travel and Tourism Council 1997a). There is, perhaps, a danger that these regional blocs will become protective of their own interests and will set up barriers to protect regional operations. But a more optimistic view is that these developments will provide the basis for a wider system of inter-regional liberalization.

There are at least eight forces which are driving the airline industry in this direction (Wheatcroft 1994). They are:

1   Increased recognition of the economic importance of tourism.
2   Pressures towards trade liberalization.
3   Moves towards privatization of airlines.
4   The marketing advantages of large airline size.
5   Moves towards mergers and alliances.
6   World-wide reservations and communication systems.
7   Increasing acceptance of foreign ownership.
8   Increased recognition of the limitations of bilaterals.

All eight of these pressures are important. They reflect major changes in the world trading system and in the structure of the airline industry. Airline privatization has proceeded to the current point at which ninety-seven of the top 194 world airlines are now fully privatized and a further eighteen have less than 50 per cent government shareholding. The airline industry is also being transformed by cross-border investments and an increasing acceptance of foreign ownership. Large airlines have been able to increase their marketing strength by controlling computer reservation systems (CRS) and communication systems. Moves towards international mergers and alliances are pointing to the emergence of four major groupings (British Airways/American Airlines, KLM/Northwest, United/Lufthansa and Swissair/Delta/SIA) that may become the 'giants' of the industry (International Air Transport Association 1996a and 1996b). But it is the first pressure, the impact of the increased recognition of the economic importance of tourism on aviation policies, which is the most important factor in the relationship between the airlines and tourism and which is the central concern of this chapter.

The objectives of tourism policy

In achieving the right balance between aviation and tourism policies, it is essential that the objectives of each should be clearly recognized. Problems arise because the objectives of tourism policy are not always precisely articulated. The objectives can be grouped into four categories: economic, social, environmental, and cultural. Sometimes there are also political objectives. And often there are conflicts between the objectives: the economic objectives may clash with the social, environmental and cultural aspirations. But in most countries the following five economic objectives are paramount:

1   Growth of national income.
2   Enhanced employment through tourism.
3   Gains in foreign exchange earnings.
4   Regional development initiatives.
5   Increasing government tax revenues.

The measurement of the economic contribution of tourism to national income is made difficult by problems in defining what constitutes the activities of tourists. As mentioned earlier, the World Travel and Tourism Council has developed a new methodology in this field and it has been collaborating with the World Tourism Organization to define tourism in a way that will enable all countries to produce their own 'satellite accounts' for tourism (see Chapter 3 in this book). This is an important activity but, for many countries, particularly those in the developing world, it is employment and foreign exchange earnings which are the most obvious measures of the economic contribution of tourism.

Current high levels of unemployment in many Western countries give the employment objective of tourism policy a high priority. This objective also has a high priority in developing countries with hidden unemployment and little scope for expansion in other activities. For all countries, the employment advantages of tourism development are first that the industry is labor intensive and, second, that it can offer many jobs for which training is rapid. Investments in tourism therefore produce a higher and faster increase in employment than equal investments in other industries (World Travel and Tourism Council 1997b).

The maximization of foreign exchange earnings is a high priority for most countries and in developing countries it is often the primary objective of tourism policy. Tourism earnings are often the most effective, or even the only way to produce the foreign exchange needed for the import of goods and services essential for national economic growth. There is, however, a danger that the pursuit of foreign exchange earnings can lead decision makers to gloss over the fact that it is *net* and not *gross* earnings which contribute to the economy. Net foreign exchange earnings take into account the imports which are needed to produce the services which the visitors require.

The primary 'leakages' of foreign exchange earnings arise from the following kinds of tourism imports:

- Imports of construction materials and equipment.
- Imports of consumable goods like food and drinks.
- Repatriation of income and profits earned by foreigners.
- Interest and repayment of foreign loans.
- Overseas promotional expenditures.

These import leakages in tourism can be very high. Various studies have shown that they can be as high as 40 per cent of total foreign exchange earnings from tourism in some countries and are rarely lower than 10 per cent (Wheatcroft 1994).

But even when the leakages of foreign exchange have been fully taken into account, the economic gains from the development of tourism can be extremely important for most countries. The major policy issue for many countries is to ensure that the potential benefits of tourism growth are not inhibited by inadequacies of air services arising from protectionist aviation policies. Such inadequacies may take the form of a shortage of capacity, poor service standards, inadequate frequency or higher than necessary fares. Indeed, the financial benefits derived from having a national airline are all too often illusory.

The costs of protectionist aviation policies

The example of tourism in a small island with a developing economy can be used to illustrate the impacts of aviation protectionism. The World Tourism Organization publication *Aviation and Tourism Policies* (Wheatcroft 1994) approached this by examining the tourism economy of a small hypothetical country called Paradise Island which had a protected national airline called Air Paradise and about 150,000 visitors a year. The example showed that the gross receipts from hotels and other tourist activities were 2.5 times as large as those of the national airline. But this was gross before taking account of export leakages. For the hotels and other tourism activities, these leakages were assessed at 31 per cent of gross earnings. But for airline operations, the export leakages are much higher because operating expenses like fuel, landing charges, crew expenses, leasing charges and many other costs are incurred in foreign currencies. As a result, the foreign exchange leakages of Air Paradise were estimated to be at least two-thirds of gross revenues. The net earnings of tourism activities were therefore shown to be five times as large as those of the national airline. Thus, it was concluded that any aviation policy, which results in capacity restrictions or higher fares, will almost certainly produce adverse economic consequences.

The major point arising from the case of Paradise Island is that many governments do not seem to be aware that such a large proportion of the revenues earned by their national airlines are spent on imported goods and services. The two-thirds figure assessed for Air Paradise is an underestimate for many small airlines. The leakages can be much higher – up to 90 per cent – when many of the staff employed by the airline are expatriates. Even in larger and more developed countries the same conclusions are valid. The foreign exchange leakages of airline operations in Australia, for example, have been estimated at 45 per cent of revenues (Wheatcroft 1994).

For more developed countries, it is the differences in employment in airlines and tourism which give the best indication of the dangers of protectionist policies. The differences can be assessed by measuring the employment created by an additional investment in hotels compared to the employment created by the same investment in a national airline. An investment of $US150 million in a new Boeing 747 can be shown to give employment to 400 persons. An equal investment in hotels can be shown to give direct employment to over 1,500 persons (Wheatcroft 1994). This is a vitally important factor when judging the merits of aviation policies.

It is sometimes argued that small countries need their own airlines in order to safeguard and guarantee the continuation of the services needed by their tourist resorts. It is argued that they cannot rely upon foreign airlines to continue to operate such services. It is true that the promotion of tourism requires the guarantee of good air service but it does not necessarily mean that those services must be operated by a protected national airline. There are other ways to achieve this policy objective which do not involve protectionist aviation policies (World Travel and Tourism Council 1997a).

Some lessons from aviation policies

Aviation and Tourism Policies presents a selection of case studies of the aviation and tourism experiences of a variety of countries in recent years. The policies examined in these case studies fall into five groups:

1   Countries where tourism has clearly benefitted from the liberalized aviation policies.
2   Countries which have maintained protectionist polices to the detriment of tourism.
3   Countries which have enjoyed a boom in visitor numbers – but not necessarily in earnings – from unrestricted charter operations.
4   Countries which have banned or restricted charter operations in order to maintain high per capita tourism earnings.
5   Countries in which policy debates continue about the best policies to adopt.

Liberalization successes

The first of these groups include countries as diverse as Australia, Mexico, Egypt, Thailand, Indonesia and New Zealand, although most of them have had their ups and downs in the pursuit of tourism expansion through aviation liberalization. Australia is the most interesting of this group of countries because the liberalization of aviation policy was quite explicitly related to a political decision that the potential gains from expanding tourism were greater than the advantages of protecting a national airline. Domestic liberalization came first, and places like Cairns enjoyed a 30 per cent growth in traffic in the first year. International tourist arrivals in Australia increased strongly after the liberalization of international aviation policy in 1992. The close alliance between Qantas and British Airways, following the decision to allow foreign ownership participation, has greatly benefitted the two airlines but they have faced increased competition from Singapore Airlines and from Britannia Airways services to Cairns. This increased competition has undoubtedly been an important factor in the growth of Australian tourism.

Thailand and Indonesia have both had very large increases in international tourist arrivals in recent years, and both are now major markets. Thailand had over six million international visitors in 1994 and Indonesia had over four million. Both countries have encouraged competitive airline operations and there are now new entrant competitors in both Thai International and Garuda markets, but both airlines continue to share in the booming tourism traffic growth of South-east Asia.

International tourist arrivals in Egypt doubled in the years from 1986 to 1990 during a period of liberalized aviation policies. Sadly, there has been little or no growth in visitor traffic since 1990 because of security concerns resulting from terrorist activities. Nevertheless, the earlier results justify including Egypt as one of the examples of successful aviation liberalization.

Tourism is very big business in Mexico. The World Travel and Tourism Council estimates gross output to be 12.7 per cent of GDP in 1996. The major international market is, of course, the United States from where almost three-quarters of visitors originate. Mexican aviation policy in the past has been very much concerned with fears about the potential dominance of US airlines in the Mexican market. Nevertheless, since 1988 there has been a progressive liberalization, which has not only benefitted the growth of tourism, but has allowed Mexican airlines to retain a reasonable market share. Mexican airlines suffered from the economic and political crisis of 1994, but Aeromexico, the largest airline, expanded its domestic and international operations by taking over some smaller airlines and investing in foreign airlines like AeroPeru.

Continuing protectionism

Most countries at least pay lip service to the notion that the liberalization of aviation regulation would be advantageous for the development of tourism. Exceptions to this are still to be found, however, particularly in parts of Latin America and Africa.

Brazil is a notable example of a country which still firmly believes in the merits of aviation protectionism. As recently as 1992 the government explicitly endorsed measures designed to protect the Brazilian civil aviation market, including the control of tariffs and the principle of predetermining capacity operated in international routes. Visitor traffic declined substantially from 1986 to 1990, and although there has been an improvement in more recent years, the traffic volume in 1994 had still not returned to that of 1986. Nevertheless, the major airline – VARIG – is recovering after restructuring and has recently strengthened its market position by joining the Star Alliance that includes United Airlines, Lufthansa, SAS, Air Canada and Thai Airlines as partners.

Protectionism also still plays a very large part in the aviation policies of many African countries and it was, in fact, the concerted opposition of African countries (with the support of Japan) which led to the failure of an ICAO conference on aviation liberalization in 1994. It had been hoped that the conference would take some preliminary steps towards relaxing international bilateral control of tariffs and capacity. But, largely because of African opposition, it concluded that disparities in economic and competitive situations ruled out the possibility, in the near future, for global multilateral agreement concerning the exchange of traffic rights. The only signs of change from protectionism have come from a realization that many African countries have neither the will nor the financial capability to continue to support their ailing national airlines. There have been one or two instances of governments, notably Zambia, allowing their state-owned airlines to go out of business. Also, there are a few bright spots – developments in Kenya, the Seychelles, Mauritius and South Africa are dealt with later. However, the general picture of air transport and tourism in Africa is not very encouraging.

Charter booms

Two countries – Turkey and the Dominican Republic – have enjoyed very high growth rates in international tourism as a result of policies that have allowed the unrestricted development of inclusive tour (IT) air services. Both of these countries rank highly in the 'tourism growth league' presented in Table 9.3.

International tourist arrivals in Turkey increased almost three-fold from 1986 to 1992, though since then there has been a small decline. The traffic boom was mainly due to the liberalization of charter operations, which grew to the point where they carried two-thirds of all air arrivals. These IT charter services were initially operated mainly by enterprises based in the United Kingdom and Germany, but subsequently a number of Turkish airlines have entered this market. The decline in visitor numbers, since the peak in 1992, may reflect a more widespread disenchantment with IT packages. Auliana Poon (1993), in *Tourism, Technology and Competitive Strategies*, presents the view that consumer preferences in western countries are changing and that what she calls 'the new tourism' will increasingly turn against packaged products (see Chapter 6 and Chapter 8 in this book). Another factor that can influence the future of the IT market is the relatively low, average daily spending of travelers on packaged tours. Countries like Turkey that have enjoyed a boom in the number of visitors are tending to view the higher daily spending of independent travelers as a target for improving the contribution that tourism makes to the national economy.

It is certainly true that IT traffic to the Dominican Republic, which doubled in numbers from 1986 to 1994, produces a very low, average daily contribution to the national economy. Average tourist expenditures have been as low as $US60 a day. The growth of tourist arrivals has been associated with an 'open skies' aviation policy, and a large number of visitors have arrived on charter flights from Canada and Europe. Furthermore, the growth has not been very well planned, and in several areas the bed capacity of hotels has exceeded the basic infrastructural provisions of water and electricity.

Charter bans

Some aspects of the experiences of boom countries like Turkey and the Dominican Republic have led other countries to conclude that they have more to gain from a smaller volume of higher-spending traffic than from the pursuit of a mass market through an unrestricted charter policy. Two such countries are the Seychelles and Mauritius.

Table 9.3 Tourism growth league[a]: increase in international tourists from 1986–94

	International tourists *1994* *(000)*	*Growth* *1986–94* *(%)*
South Africa	3,897	500
Macau	4,489	483
Indonesia	4,006	383
Vietnam	1,018	308[b]
Zimbabwe	1,139	206
China	5,182	196[b]
Turkey	6,034	189
Chile	1,623	179
Hong Kong	9,331	150
Cyprus	2,069	149
Australia	3,362	135
Dominican Republic	1,717	129
Malaysia	7,197	124
Argentina	3,866	119
Thailand	6,166	119
Singapore	6,268	116
Korea	3,580	116
Malta	1,176	107
Hungary	21,425	102
Egypt	2,356	90
Uruguay	1,884	89
Philippines	1,414	86
Japan	1,915	81
New Zealand	1,323	81
USA	45,504	79
Puerto Rico	3,042	79
Ireland	4,309	74
Norway	2,830	73
France	61,312	69
Portugal	9,132	69
Israel	1,839	67
Colombia	1,207	66
Morocco	3,465	63

Source: *WTO Compendium of Tourism Statistics, 1990–94.*
Notes:
a Countries with more than one million international tourists in 1994 which exceeded world average growth of 60 per cent from 1986–94.
b Growth 1990–94.

The Seychelles has enjoyed a steady, but not spectacular, increase in tourist visitors for many years, and this reflects a governmental policy of controlled growth and a charter flight ban. A government policy statement in 1993 said that 'charter operations will only be permitted when they do not adversely affect the viability of scheduled services' (quoted in Wheatcroft 1994: 71). As a result, very few charters have been authorized. The Seychelles aviation policy has also been based on the belief that tourism interests are best served by having a national airline that is totally committed to the long-term development of traffic to the island. The national airline, Air Seychelles, has been supported, but not subsidized, as part of this policy.

There is little doubt that the ban on IT charter operations has resulted in a much higher average daily tourist expenditure in the Seychelles than that achieved in countries which have experienced a charter boom, and this is the justification for a policy of restricted growth. Nevertheless, tourism businesses in the Seychelles exhibit many of the characteristics of the hypothetical Paradise Island described earlier, particularly in the high level of foreign exchange leakages, both in tourism activities and even more so in the operations of the national airline (Wheatcroft 1994). It must, therefore, be questioned whether the protectionist policy of the Seychelles government will be beneficial in the long run. An issue to be considered is whether a greater contribution to the island's economy could be achieved by a more liberal aviation policy, with foreign airlines bringing a larger volume of traffic to the country, and thus reducing the foreign exchange expenditures of providing air services.

In many ways, the aviation and tourism policies of Mauritius are similar to those of the Seychelles, particularly in the ban on charter operations and the pursuit of high-spending visitors. But Mauritius has a much larger tourist industry and much more ambitious plans for future growth than the Seychelles. It enjoyed a 143 per cent growth in visitor traffic from 1986 to 1994, which would have put it high in the growth league table if it had met the size qualification. Virtually all visitors to Mauritius come by air and all on scheduled air services. The island is well served by a network of services linking it to its major markets in Europe and South Africa. But price competition on these services is regulated and air fares tend to be a high percentage of the total cost of a holiday visit. The problem for the Mauritius government is that it wants to maintain the high-quality image of the island as a tourism destination and the high average daily rate of spending of its visitors, even though it has allowed an over-expansion of tourist accommodation and created internal pressures for a relaxation of aviation regulation to create more traffic.

Continuing debates about aviation and tourism policies

There are quite a number of countries in which debates continue about the best balance of aviation and tourism policies. Among these, some of the most interesting are Kenya, India, Israel, Cyprus and South Africa. Each of these has important tourism industries and each has made at least some moves towards more liberal aviation policies for the benefit of tourism development.

Until recently, the aviation policy of Kenya was essentially protectionist with the seat capacity on scheduled services being 'predetermined' under bilateral agreements and controlled for the benefit of the national airline. The authorization of foreign charter services, albeit only to Mombasa and not to Nairobi, created a new competitive situation, and traffic on the scheduled services of Kenya Airways remained static for several years. A major recent change in the aviation policy has been the decision to privatize Kenya Airways and, under a new foreign management, the operations and financial results of the airline have greatly improved. There is a reasonable expectation that this change in the status of the national airline will lead to the relaxation of protectionist aviation policies, and that the tourist industry will benefit from these changes.

India is, without doubt, the sleeping giant of the world's tourism industry. In relation to the size of the country and the wealth of tourism attractions available, it has failed, in the past, to realize its huge potential. Less than two million tourists visited India in 1994 compared with the six to seven million tourists who visited Malaysia and Thailand. Moreover, the 30 per cent visitor growth rate in India from 1986 to 1994 is very low compared with the world average of 60 per cent.

There can be little doubt that restrictive aviation policies have contributed significantly to the under-development of tourism in India. Past policies have been restrictive and designed to protect the commercial interests of Air India. Some important changes have been made in the past two years and domestic operations have been opened up to competition to the extent that new entrant airlines carried over one-third of the traffic in 1995. Some new competitive operations have also been allowed on international routes. The potential exists for the privatization of Air India and Indian Airlines, but although this has been talked about for several years, there does not seem to be an immediate prospect of it being implemented. India continues to debate the advantages for tourism of a less regulated, air transportation market place.

Israel is yet another country in which the advantages of aviation liberalization continue to be debated. In 1994 the government accepted the recommendations of an inter-ministerial committee which recommended some major changes including the designation of an alternative competitive airline to EL AL, and the relaxation of predeterministic capacity regulation on international flights. The government also reaffirmed its objective of privatizing EL AL but this has not yet been implemented. These changes, together with more liberal rules for the regulation of charter flights, offer considerable new advantages for tourism development. Moreover, progress towards peace in the Middle East could also bring new opportunities. A new joint airport servicing Eilat and Aqaba has the potential to bring great benefits to the tourism markets of both Israel and Jordan.

Cyprus has enjoyed a high visitor-growth-rate and the industry has made an important contribution to the economy of the island (Ioannides 1992 and 1994). Much of the increase in tourist traffic has come from IT charters operated by foreign airlines. Cyprus Airways, the national airline, has tried to compete with these foreign charters by setting up its own charter airline subsidiary but this has not been too successful. Recent policy debates in Cyprus have centered on whether there should be more restrictions on foreign charters to protect the operations and finances of Cyprus Airways. Some support for such policies has come from hoteliers and other sectors of the tourist industry; they have become concerned that a weakening of the position of Cyprus Airways would leave them entirely dependent on foreign tour operators who would therefore be able to demand even lower prices for services provided. This is a real concern because, like other countries highly dependent on IT charter operations, the average daily expenditure of visitors to Cyprus is already fairly low.

South Africa heads the list of countries in the tourism growth league (Table 9.4) with a five-fold increase in traffic from 1986 to 1994. Radical political changes and the end of sanctions have totally transformed the prospects for tourism in this country. Visitor numbers increased by 15 per cent in 1994 and, with the additional stimulus of the Rugby World Cup, tourist traffic increased by over 50 per cent in 1995. The more liberal aviation policies which have been introduced since 1992, in both domestic and international aviation regulation, have played an important part in this transformation. In the international field, South Africa has declared that its future policy in the negotiation of bilateral agreements will allow the multi-designation of airlines, will eliminate capacity restrictions and will abolish tariff controls. These are major steps towards a liberated market. The number of international airlines serving South Africa has doubled since 1990, and services are available to many more countries.

South Africa has said that it intends to retain controls on charter operations so that they are no more than 10 per cent of scheduled flights. The declared objective of these restrictions is to ensure that South Africa does not become a cheap charter destination but promotes itself as a quality destination. A survey of departing air passengers in 1994 reported an average daily expenditure of $US51. This is relatively low compared with the Seychelles, for example, where the daily expenditure is over $US100. On the other hand, the average length of visit in South Africa is high, ranging from twenty-four days for VFR (visiting friends and relatives) to seventeen days for business visitors, with an average of twenty days for all visitors. It will be interesting to see what effects the restriction on charters has on future traffic.

In the domestic aviation field, the government deregulated air services in 1992 and the immediate effect was to stimulate competitive operations. Subsequently, however, there have been a series of failures and mergers which have re-established the dominant position of South African Airways (SAA) in the domestic market. It is too early to forecast the future pattern of competitive services.

One other aspect of aviation policy development in South Africa is worth noting. The new South Africa is beginning to take a leading role in the development of regional policies for Southern Africa. By joining the Southern African Development Council, South Africa is committed with other countries of the region to work together in the negotiation of future air service agreements. Equally important is the African Joint Venture in which three airlines – Uganda Airlines, Air Tanzania and SAA – have created a joint venture airline to be known as 'Alliance' which will operate initially with SAA aircraft and crews. It is this kind of cooperative activity which can create regional airlines capable of operating without subsidy or protection.

Air services and tourism planning

Some important conclusions emerge from the foregoing analysis of the potential conflicts between aviation and tourism policies. The first is that the provision of an adequate system of air services is an essential requirement for the successful development of tourism to very many destinations. This may seem obvious but it is curious that so many planning studies for resort development ignore or greatly underestimate this basic fact; they give inadequate attention to the network of air services needed to exploit the major traffic generating markets and to the CRS and other marketing facilities essential for the success of these operations. It is not yet fully recognized that airlines have tremendous power in determining the success or failure of tourism developments through the manipulation of origin–destination tourist flows.

The techniques of tourism planning have made great advances in the past decade, and much emphasis is rightly put on an integrated analysis of all the elements which are needed to achieve sustainable growth. What is often missing, however, is a recognition of the vital importance of efficient air services in this process and, in particular, the need to adapt aviation regulatory policies to the needs of tourism.

There are various reasons for this neglect of the importance of air services in tourism planning and policymaking. One is that the data required for the effective evaluation of alternative aviation and tourism policies is often far from adequate. Information about foreign exchange leakages, for example, is difficult to obtain and does not appear to have been properly evaluated in most countries. This is true for tourism activities like accommodation, catering and entertainments and, most particularly, for airline operations.

Very few developing countries have recognized the low level of the net revenues that they are earning from the operations of their national airline, despite the fact that one of the main justifications for sustaining such an airline is that it earns foreign currency.

Other aspects of the travel and tourism industry are also inadequately researched and much more information is needed about the contribution of the industry to national economies. The collaboration between the World Travel and Tourism Council and the World Tourism Organization to agree on standard principles for national satellite accounts for tourism is a major innovation in this field. Such a development should increase governmental understanding of the economic importance and structure of the travel industry. While new data will be essential for the evaluation of alternative policies, the data collected so far has led the WTTC and the WTO to conclude that more liberal aviation policies are beneficial in increasing the economic contributions of tourism. Indeed, aviation liberalization has been adopted by the WTTC as an important objective within its overall approach to promoting more open and competitive tourism markets.

In those countries where the arguments for aviation liberalization have not yet won the day, it is often true that the influence of those concerned with protecting the national airline is greater than that of other sectors of the travel and tourism industry. This is partly due to the fragmented nature of the industry, and the lack of a single voice to speak for the interests of hotels, restaurants and entertainment activities. It is noteworthy that in countries in which a single organization has been formed to bring together these different parts of the industry, there has been a significant change in the balance of aviation and tourism lobbies in policymaking. It was noted earlier that a material factor in the development of a liberalized aviation policy in Australia was the clear articulation of a coordinated policy for tourism. This could happen elsewhere.

Another important factor in the balance between aviation and tourism interests is the structure of government. It has been generally true, in the past, that aviation has had a higher level of representation in government than tourism. And national airlines, particularly when state-owned, have been able to exert strong pressures on government policies. Tourism has had a lower profile in government organization and this, combined with the fragmented nature of the industry, has led to a much weaker role in shaping policy-making. Fortunately this is beginning to change, and the relative profiles of aviation and tourism ministers are becoming more equivalent or are even being merged into a single ministry. It is to be hoped that this will eventually be reflected in the weights given to the economic contributions of the two sectors, and will increasingly lead to the adoption of more liberal aviation policies.

It should not be concluded from all that has been said about future developments in the airline industry that there will be no place for small national airlines. It is true that the most likely scenario for the future is that the airline industry will be dominated by a small number of very large multinational airlines with global networks. But there will continue to be a place for smaller airlines operating as niche carriers in specialized fields. However, it is important for the development of tourism that such carriers should prosper on their own merits and should not have special protection from foreign airline competition.

10

CONTINUITY AND CHANGE IN THE HOTEL SECTOR

Some evidence from Montreal

Simon Milne and Corinne Pohlmann

Introduction

In 1992 Meric Gertler coined the term 'forgotten sectors' to draw attention to the fact that economic geographers were largely ignoring important changes occurring outside the manufacturing arena. This 'manufacturing bias' is now showing signs of waning. Several excellent reviews of service industry trends, locational dynamics, and related theoretical issues have emerged in the literature in recent years (Christopherson 1989; Christopherson and Noyelle 1992; P. Wood 1991a and 1991b; Daniels 1991; Allen 1992; Begg 1993; Coffey 1995). While this body of work is quite diverse in content, one common theme to emerge is that geographers need to adopt a more 'services informed' view of current economic and societal change (Marshall and Wood 1992: 1,264). Such a view must, in part, be based on a thorough understanding of the patterns of production organization and labour utilization that characterize tertiary activities. At the same time, there is a need to look more closely at how services are represented in studies of regional economic restructuring and, in particular, how their relationship to other sectors of the economy, including manufacturing, is evolving (Urry 1987; Townsend 1991; *The Economist* 1993a; Milne *et al.* 1994).

A number of factors have hindered our ability to generate a more 'services informed' approach to understanding current economic change in Canada. While there has been a growth in empirical research on the country's service sector, the bulk of the work by geographers has tended to focus on a rather narrow array of activities, especially business and/or financial services (Coffey 1992: 1995). Relatively little attention has been paid to the diverse and over-lapping geographies of public and consumer-oriented services (Picot 1986; Milne *et al.* 1997). This is an important oversight as the latter two categories made up 59 per cent of the country's real GDP generated by tertiary activities in the early 1990s, and employed over 65 per cent of the total service-sector work-force (Industry Science and Technology Canada 1991; Economic Council of Canada 1990).

In this chapter we aim to strengthen our understanding of the evolving structure of consumer services, and how these shifts are likely to influence certain elements of urban economic systems, by focusing on the Montreal hotel industry. We begin with a brief review of the important role played by hotels in metropolitan areas and outline their particular significance to Montreal. We then turn our attention to the competitive environment facing the city's hotel industry in the early 1990s. Changing consumer demand, market saturation, globalization, and macro-economic pressures are shown to be just some of the factors that were forcing a re-evaluation of existing managerial and organizational strategies. In analysing corporate reorganization we focus on three themes: labour management; the adoption of new technologies; and the development of strategic alliances and networks.

Our discussion then shifts to the regional economic implications of the findings. We show that certain groupings of hotels are more likely to survive the competitive pressures of the 1990s than others, with non-specialized, medium to small, independent operations being most at risk. We also point to a number of labour market implications that stem from the current management practices being adopted by hotels. The conclusion stresses that the complex nature of the service being provided by hotels, and the broad range of corporate restructuring approaches being adopted, make it difficult to generalize about an overall 'trajectory of change' for this particular consumer service sector. It is certainly difficult to talk of a full-blown shift from old (Fordist) to new (post-Fordist) forms of organization, indeed we are witnessing processes of both continuity and change. If researchers are to understand better the evolution of the service sector, and its impacts on local economies, we need to adopt a more broad-based and nuanced approach – one that embodies consumer as well as producer service perspectives, and one which does not rely too unquestioningly on manufacturing-biased frameworks.

Much of our analysis is based on semi-structured interviews (1–2 hours in

Table 10.1 Basic characteristics of the Montreal hotel sample, 1992

| | *Number of hotels by ownership type* | | |
Size[a]	*Independent*	*Quebec-based chain*	*Non-Quebec chain*
Small (1–60 rooms)	22	–	–
Medium (61–229 rooms)	11	3	11
Large (230+ rooms)	1	2	11

Source: Based on Montreal hotel industry survey.
Note:
a Hotel sizes ranged from three rooms to 1,030.

duration) conducted with the general managers of sixty-one Montreal hotels during 1992–93 (see Pohlmann 1994) (Table 10.1). The sample represented approximately one-quarter of all hotel operations in the greater Montreal area. Most of those enterprises interviewed were located in the downtown core of the city. A small number of large and medium-sized establishments situated at the city's busiest airport (Dorval) were also included. Additional interviews were also conducted during 1992 and 1993 with relevant hotel associations. Our discussions on alliance formation and the adoption of reservation technology are supplemented by in-depth interviews conducted with a smaller group (15) of downtown hotels in 1994 and 1995 (see Gill 1997).

The importance of the hotel industry

Hotels are vital components of urban tourism products and make important contributions to both national and local economies. Accommodation services are estimated to account for over 15 per cent of total tourist expenditure in Canada, with hotel-based food and beverage activities accounting for an indeterminate and considerably smaller amount (Tourism Canada 1990). In 1991 the accommodation sector employed approximately 150,000 Canadians on a full or part-time basis (1–1.5 per cent of the nation's work-force). Women were estimated to make up 65 per cent of the total labour force, and 70 per cent of all the part-time workers (Statistics Canada 1992). The industry also generates a range of downstream benefits for local economies; indeed a study of the downstream impact of various service industries on Canadian-goods producers found accommodation and food services to be among the top performers (Economic Council of Canada 1990: 8).

While accurate figures on the significance of hotels to the Montreal economy are not available, the forty-four large and medium-sized establishments that make up the Hotel Association of Greater Montreal (HAGM) generated $403.5 million in revenue during 1990 (65 per cent of which was spent by visitors from outside Quebec). Association members also employed the equivalent of 7,715 full-time workers, paying $202 million in salaries and other employee benefits (source: interviews). The relative importance of hotels as urban employers has been exacerbated by the erosion of the manufacturing base of most North American cities since the 1970s (Noyelle and Stanback 1984; Levine 1989; Fainstein 1990). In Montreal, for example, over 80 per cent of the urban labour force is now employed in services; a dramatic increase from the 59 per cent who worked in the sector in 1971 (Ville de Montreal 1993a and 1993b; Pohlmann 1994).

The importance of hotels to urban areas cannot, however, be seen simply in terms of employment and revenue generation. As the base for the bulk of overnight visitors to a city, hotels exert considerable influence over the competitiveness and nature of urban tourist products. The degree to which hotels are able to cater to convention business needs will, for example, be an important factor in determining the ability of a city to out-compete other urban areas in this highly competitive market. At the same time, hotels are vital information disseminators, influencing visitor perceptions of, and actions in, the surrounding environment through in-room 'what-to-do' packages and TV displays.

Hotels also play a pivotal role in attempts to regenerate urban cores, and recreate heritage sites, as cities attempt to reverse the ravages of deindustrialization and restructure their economic base (see Harvey 1989a: 88–92; S. Britton 1991: 467; Watson 1991; Chang *et al.* 1996). Hotels often form a vital part of the cultural, capital-driven complexes that are so important in transforming existing built environments – altering traditional economic roles, and enabling cities to forge distinctive images (Zukin 1995; Tufts and Milne 1997).

The evolving competitive environment

At the time that most of the hotel interviews took place (summer 1992, winter and spring 1993), Montreal was struggling to emerge from a severe economic down-turn. Recessionary pressures in 1991 led to slower growth in leisure and business travel in Canada than in previous years. Between January and June of 1991, the number of trips by non-residents was down 2 per cent compared to the previous year, with the usually strong month of July seeing a 3 per cent decline. Quebec saw an overall decline in tourism business of 10–15 per cent compared to previous years (Tourism Canada 1991). Business travel also fell as many firms attempted to trim their budgets. During the first six months of 1990 there was a decline of 50,000 room nights in the Montreal urban area compared to the same period for 1989. The average occupancy rate for the city's hotels stood at 61 per cent for the year – down from 67 per cent in 1987. In 1991 occupancy rates fell further to 55 per cent (Table 10.2). Montreal and its other major Canadian competitor, Toronto, also had to face a significant slump in the intensely competitive conventions market (Purdie 1992).

Table 10.2 The evolution of Montreal hotel industry occupancy rates

Year	Available room nights (000)	Actual room nights (000)	Occupancy rate (%)
1979	3,892	2,650	68.1
1983	4,256	2,423	56.9
1987	4,448	3,082	69.3
1991	4,916	2,717	55.3
1995	4,728	2,947	63.8

Source: Hotel Association of Greater Montreal

Twelve of the small hotels interviewed (55 per cent) had registered falling occupancy rates. Interestingly, several of the other small hotel managers felt that the recession had boosted business, with travellers perceiving them as relatively cheap alternatives to larger operations. Nearly 70 per cent of the medium-sized hotels had seen occupancy rates drop. The trend among the large hotels was similar, with ten of the fourteen noting a down-turn in occupancy. Medium sized, independent hotels, were feeling recessionary pressures most acutely. Large chains, with their relatively deep pockets, and small operators with limited, fixed-wage costs, appeared to be best able to lower prices and ride out the recession.

While the impacts of recessionary pressures did not vary greatly according to the location of hotels, airport based companies did, in most instances, suffer more than their counterparts based in the central city. It appears that this variation was largely due to the former's heavier reliance on the business travel sector which had been more adversely affected by the economic down-turn than leisure-focused tourism.

The city was also preparing for the possibility of another referendum on Quebec's relationship with Canada. Several of those interviewed were scathing about the 'language issue' and the unstable political reality of Quebec, claiming that these factors scared off US tourists and also discouraged inward investment. On some occasions managers of hotels with a heavy reliance on the business sector also spoke of the down-turn in clients that would occur if established firms decided to leave an independent Quebec.

Consumer demand

Over 85 per cent of those surveyed felt that consumers were demanding more value for money than in the late 1980s. The growing demand for value stems, in part, from reduced corporate and individual spending due to recessionary factors, but is also generated by the wholesalers and tour agents/brokers who account for approximately one-third of the room-nights booked annually in Montreal (source: interviews). These buyers are using increasingly sophisticated technologies to compare prices and find alternative hotel space. They are able to play one hotel off against another, and push prices down, while also demanding higher quality and a greater variety of services (Go 1992b: 23). As a result, hotel sales staff must be prepared to do more negotiating and bargaining, and effective-yield management systems become increasingly important.

Demographic factors were also influencing consumer demand in profound ways, with several managers highlighting relatively significant growth in the older-age travel market (see also McDougall and Davis 1991). Most (67 per cent) of the small hotel managers interviewed felt that the number of seniors in their clientele had been stable, with some claiming that older travellers were not as attracted to smaller operations because they often could not provide the conveniences, such as elevators, found in larger establishments. Over 90 per cent of medium-sized hotels had witnessed growth in the seniors market with similar figures emerging for large hotels. Some managers of larger hotels also mentioned the fact that they had been forced to introduce changes to cater for the growing number of female business travellers coming to Montreal, including the creation of separate exercise rooms.

Market saturation and ownership concentration

The economic boom of the 1980s led to considerable hotel construction and, by the 1990s' recession, cities in much of Canada were generally over-supplied with hotel rooms (Waldie 1993). The increasing saturation of domestic markets has made it difficult for independent hotels to obtain a return on investment. At the same time, larger international hotel chains are attempting to increase profits and market share by investing overseas, creating global competitive pressures (Dunning and McQueen 1982a; Laventhol and Horwath 1989; Littlejohn and Beattie 1992; Gannon and Johnson 1995).

Montreal was no exception to these trends. The number of hotel rooms in the city grew dramatically during the 1980s and occupancy rates also increased. Much of this growth was fuelled by new building by multinational chains. The recessionary period saw some of the lowest occupancy rates in the city's recent history, and by 1995 the city's accommodation capacity had shrunk by about 5 per cent from 1991 levels (Table 10.2).

Perceptions of the market saturation issue varied considerably according to the size and ownership profile of the hotels. Over-supply of rooms was rarely mentioned as a problem by the twenty-two small hotels surveyed in 1992. Indeed, a small number of interviewees said they felt there was more room for high-end, boutique-style, hotel development in the city. On the other hand, almost all managers of medium and large operations were concerned by the prospect of additional accommodation construction in the city. Several managers specifically stressed that any further building by multinational chain hotels in the urban core would be disastrous for the existing industry.

Chains have increased their overall share of global accommodation capacity considerably in recent decades, with the total number of rooms controlled by the world's twenty-five largest companies virtually doubling every decade since the 1970s: 712,000 in 1970; 1.32 million in 1980; and 2.45 million in 1990 (*Hotels* 1992: 15) (see Table 10.3). While accurate figures are not available on the changing role of chains in Montreal's overall accommodation profile, our respondents were in little doubt that the relative importance of these operations had grown during the past two decades. While US-based companies are important among Montreal's chain operations, Canadian and European companies are also present. The strong US involvement is not surprising when one considers that seventy of the world's largest 200 chains,

Table 10.3 The world's ten largest hotel companies, 1995

Chain (HQ location)	Rooms	Number of hotels	Hotels added 1994–5
Hospitality Franchise Systems (USA)[a]	509,500	5,430	1,139
Holiday Inns Inc. (USA)	369,738	2,096	166
Best Western International (USA)	282,062	3,462	53
Accor (France)	268,256	2,378	113
Choice Hotels International (USA)[b]	249,926	2,902	75
Marriott Corp. (USA)	198,000	976	125
ITT Sheraton Corp. (USA)	129,201	414	(11)
Hilton Hotels Corp. (USA)	90,879	219	(7)
Promus Companies[c] (USA)	88,117	669	99
Carlson/Radisson/Colony (USA)	84,607	383	34

Source: *Hotels* (1996: 46).
Notes:
a Includes Ramada, Days Inns and Howard Johnson brands.
b Includes Quality, Sleep Inn, Econo Lodge and Comfort brands.
c Specializes in Casino hotels.

and nine of the top ten, are based in the US (*Hotels* 1996) (Table 10.3). It is also important to note here that not all investment and development by chains involves the construction of new establishments. Franchising deals, and a growth in the use of management contracts, have also enabled chain operations to add hotels to their groups.

Hotel restructuring

Our research revealed that hotels were adopting a broad range of responses to the evolving competitive environment of the early 1990s. We now highlight three broad areas of corporate response: labour management; the use of evolving technologies, and the formation of networks and strategic alliances. The current processes of restructuring occurring in these areas are shown to have important implications for urban areas: implications that will be felt for some time to come.

The labour dimension

Before proceeding to our empirical findings, it is useful to first review some key characteristics of the hotel workplace. Medium and large hotels are often organized along relatively rigid, departmental lines (food and beverage, accommodation services, security, etc.) with some departments boasting a relatively wide range of job types (R. Wood 1992). The diverse range of positions found within any one hotel – from bed making to banquet preparation – will clearly make it difficult to discern any dominant 'labour process' trends that may be emerging. This complexity is only heightened by the fact that the labour force comprises two broader categories which cut across these divisions. *Front-line* workers are involved in direct contact with the consumer (waiters, front office staff, etc.), while *background* workers perform 'behind the scenes' tasks (making beds, preparing food, accounting, etc.) (see Drucker 1992).

Performance in background jobs is largely measured in terms of quantity (how many bedrooms can be tidied during a shift), while quality is largely a matter of meeting externally imposed criteria. The performance of a front-line worker embodies both quantity and quality, with behaviour toward customers often viewed as being just as important as the physical labour undertaken in performing the task (Drucker 1992; Urry 1990: 40; S. Britton 1991).

Technology, sub-contracting and lay-offs

In his detailed survey of labour use in hotels, R. Wood (1992: 133–37) points to some common misconceptions: that the industry is simply labour inten-sive; that technology can play only a limited role in improving labour productivity and reducing costs; and that, where technology is introduced, it is inevitably associated with processes of de-skilling. The hotel industry is, in fact, *both* labour and capital intensive (depending on the job in question) and is increasingly technology driven. Front-office and back-office duties, clean-ing and room preparation, security, maintenance, laundry, and restaurant/bar services are all experiencing major technological change (Tourism Canada 1988; Gamble 1991). Computer management systems are spreading rapidly through larger hotels with the benefits reported including increased staff productivity and decreased operating costs (Pohlmann 1994).

Nevertheless, the cost of new technology is often high, and for many operators (both small and large) there is little sense in replacing inexpensive (and flexible) human labour with relatively inflexible (and often expensive) automation. In this respect, the theme of work-place reorganisation and the use of 'customer labour' becomes paramount. In the kitchen, it is the menu itself that provides management with the greatest opportunity to reduce labour requirements: a move toward simpler foods can allow substantial cuts and de-skilling among cooking staff, while self-serve buffets can allow fewer serving staff to be used. Similarly, room service can be scaled back through the introduction of mini-bars or coffee machines. Front-office activities have also been affected by technology, especially the advent of in-room, check-out facilities on TV and guest-room faxes (see Grimes 1991). By reducing or simplifying certain tasks, technology can allow front-line workers to focus more on the people-oriented activities that are so critical to a hotel's success (Quinn and Paquette 1990: 70; Schlesinger and Heskett 1991). Technology seems likely, therefore, to raise the overall skill profile of the hotel work-force by reducing, in absolute terms, the number of unskilled or repetitive background functions, and broadening the customer contact functions and technological skill requirements of front-line workers.

Only three of the twenty-two small Montreal hotels surveyed used computerized systems to run day-to-day business activities, although a larger number had stand-alone personal computers for accounting purposes. Nearly three-quarters of the medium-sized operations had computerized operations while all of the large firms were automated, with the exception of a privately run 'boutique' enterprise. The consensus among many small hoteliers was that computerization is relatively unimportant to their competitiveness and that they would rather focus attention on human-contact dimensions, a finding replicated in later studies (Gill 1997). The managers of some smaller operations also complained that it was difficult to find affordable, customized software to meet small hotel needs. On a few occasions, interviewees also admitted to quite simply being 'techno-phobes'. Over 90 per cent of the small hotel managers interviewed stated that they had never laid-off workers as a result of technological change, and two-thirds felt that future technology-related lay-offs would be limited. This contrasts dramatically with the 75 per cent of large hotel managers who felt that the introduction of technology had effectively eliminated some positions in the past decade. Middle management positions appeared to have been most affected by the interrelated processes of technological change and corporate downsizing. It should be noted, however, that in some instances technology has also played a role in creating new positions. For example, the increased amounts of data and information created by computerized reservation and yield management systems, has sometimes created the need to hire workers to analyse and interpret data.

While subcontracting represents another way for hotels to reduce labour costs, it is by no means a new strategy, and its levels appeared to have increased little in Montreal during the decade prior to the interviews. Approximately 60 per cent of those surveyed subcontracted some of their operations – with laundry and maintenance being the most common areas. Half of the small hotels had adopted some form of subcontracting, while for the medium-sized hotels the proportion was greater (68 per cent). The lower level of this activity among large hotels (57 per cent) reflects their tendency to be more heavily unionized, with resultant restrictions on the range of activities that can be outsourced. Indeed several managers of large hotels said that they would seriously consider expanding the use of subcontracting if union resistance diminished.

The high proportion of establishments involved in some form of sub-contracting is yet further evidence of the powerful role that consumer services can play in generating downstream economic benefits for urban areas. In addition to supporting a large number of laundries and food and equipment supply operations, the industry also helps to support a number of high-end producer services including accounting and consulting firms, and small software companies.

Labour flexibility

Much of the literature dealing with hotel labour-use stresses the casual, non-unionized, and highly feminized nature of the work-force, and the fact that it is characterized by high levels of turnover (S. Britton 1991; see R. Wood 1992). An overview of the changes in the nature of the Canadian accommodation work-force between 1971 and 1991 reveals that, while this view of hotel labour is in many ways correct, the situation is more complex than we might expect (Statistics Canada 1988 and 1992). While basic pay rates remained low during the period (diminishing in relation to average industrial incomes), and the dependence of the industry on female labour increased, we also witnessed a relative reduction in the use of part-time labour and a significant improvement in the educational profile of the work-force. The number of young people entering the hotel industry appears to be falling in relative terms, and many of the growing number of older women in the industry are breadwinners and/or single parents. Thus, many accommodation workers are adults with families, and it is highly likely that the work they do is permanent in nature (see Schlesinger and Heskett 1991).

The use of part-time and seasonal labour by our respondents was not a new phenomenon and, in keeping with national sectoral trends, had actually declined slightly in the years prior to the survey. Just over one-third of the hotels surveyed would normally cut their casual staff by at least 50 per cent during the low season; a further 39 per cent would cut fewer than half of their staff. Only one quarter of hotels maintained full employment levels throughout the year. There was little variation according to hotel size.

The average turnover rate for all hotels hovered around the 10 per cent level, with levels higher for casual rather than permanent workers. Increasingly management is realising that high levels of turnover not only increase things like training costs and worker-compensation-claim rates, but also have a negative impact on consumer service, levels of return business, and the ability of hotels to improve levels of functional flexibility (Schlesinger and Heskett 1991: 76). Several managers felt that the low pay offered by the industry as a whole was a major factor in boosting turnover levels and reducing worker performance; at the same time, however, they recognized that basic wage rates were unlikely to rise significantly because of increased cost-cutting pressures. Nevertheless, a number of initiatives were already in place to bolster worker performance. For example, close to 80 per cent of the large hotels and 65 per cent of the medium-sized operations had set up formal employee incentive programs, often revolving around 'employee of the month' schemes. Such approaches were less commonly adopted by smaller hotels.

Management in many hotels were also making considerable efforts in the area of employee training. The driving forces here were the need to boost service levels and quality, while preparing workers for the new tasks that accompany technological change and work-force reductions. None of the small hotels interviewed operated a formal training programme, with most management feeling that such an approach was unnecessary given their small size. Over 40 per cent of the medium-sized hotels had formal training programs, while nearly 80 per cent of the large hotels had instigated such approaches. International chain hotels were the most likely to provide detailed training programs. On rare occasions management had also instituted 'quality circle' approaches, with employees and management discussing possible improvements in specific areas of hotel operation.

The gender dimension

Low skill (and low pay) positions (e.g. housekeeping) in the Montreal hotel industry tend to be heavily reliant on female employees. Women also tend to be more heavily represented among part-time and seasonal positions in general (see Bagguley 1990). Despite these trends, the proportion of women found at the management level was relatively high, with 50 per cent of the hotels surveyed claiming that at least half of their management personnel were female. The majority of this employment, however, occurred in the middle and lower management areas, with a 'glass ceiling' effect appearing to hinder upward mobility (see Guerrier 1986; Hicks 1990). Over one-third of the small hotels surveyed were run by women, but among the medium-sized hotels the figure fell to 24 per cent. No hotel larger than 200 rooms was managed by a woman.

Unionization

Twenty-three of the sixty-one hotels interviewed were unionized (5 per cent in small hotels, 32 per cent in medium hotels, and 100 per cent in large hotels). The high level of unionization in large operations was often a concern to management. There were worries that unions would discourage workers from accepting 'multi-tasking' initiatives, technology-related redundancies, and rather limited pay increases. Management also felt that several non-service-related unions, that had entered the hotel sector in search of new members, did not always have a true understanding of the difficult environment facing the industry. The rather tense relationship that appears to exist between management and organized labour in some hotels was borne out by the fact that, relatively soon after the interviews were completed, two of the larger hotels faced strike action.

Product development and computer reservation systems

Sophisticated market segmentation, and diversified product development, have been key competitive strategies in the global hotel industry for some time now (Horwath and Horwath 1987; Laventhol and Horwath 1989). Recent years have seen a proliferation of mega-resorts, all-suites, extended-stay complexes and budget hotels (Leitch 1989; Welihan and Chon 1991; Pohlmann 1994). The growth of budget hotels and all-suites has been particularly important in Montreal, with three-quarters of the managers interviewed feeling that there had been a growth in these types of establishments and that they had a good competitive future.

The hotel industry is also strongly influenced by the development of sophisticated reservation systems which allow travel agents and individuals to check pricing, construct travel packages, and make bookings through a computerized information network (World Tourism Organization 1991). Large chains operate their own computerized reservations systems (CRSs) which are, in turn, linked to airline CRSs through switching systems (A. Beaver 1992; Lindsay 1992a; *Hotels* 1996). The cost of such links is high: existing in-house reservation systems must be upgraded, while the commissions involved may, in some cases, represent 20–30 per cent of advertised room rates (Go 1992b; McGuffie 1990). Despite these costs, the added market-place profile afforded by such systems cannot be denied (see Chapter 7 in this book). CRS also provides companies with detailed customer profile data, allowing them to direct their sales to specific client bases. The systems can also facilitate the development of sophisticated yield management approaches, allowing companies to match pricing structures to seasonal and weekly fluctuations in demand, while also maximizing occupancy rates.

The increasing dominance of CRS as a marketing and distribution tool has a number of important implications for hotels. If operations are not listed on these systems they run the risk of being cut out of the market-place (see Chapter 7 in this book). For independent hotels and smaller operations, the cost of promoting themselves through CRS can rarely be borne alone, and as a result these operations face an up-hill battle in gaining access to customers. Only two of the small hotels interviewed in 1992–93 were part of a CRS. Nearly two-thirds of the twenty-five medium-sized hotels were linked to a CRS, and all but one of the large hotels was connected (Table 10.4). Most of those surveyed felt that the need to have some form of access to a CRS would increase for large and medium-sized firms in the future. There was greater uncertainty about the future value of such links for small hotels, however, with many operators feeling they were simply too small to ever need such access. A more detailed study of hotels, CRS, and networking in Montreal and Toronto revealed similar trends (Gill 1997). Among the thirty-six cases chosen from the two cities the researcher found that small, independent hotels were dramatically under-represented on global distribution systems and that this lack of representation created important competitive disadvantages.

Table 10.4 Variations in Montreal hotel CRS-use by size and ownership type

Hotel	Small		Medium		Large	
	CRS	No CRS	CRS	No CRS	CRS	No CRS
Independent.	2	20	4	7	–	1
Quebec chain	–	–	2	1	2	–
International chain	–	–	11	–	11	–

Source: Based on Montreal hotel industry survey.

In an attempt to improve the ability of small tourism operators to reach the market-place, Tourism Quebec decided to develop a regional reservation system (RRS) called Reservations Quebec (RQ) at a cost to investors of $14 million (with a $5 million guarantee by Tourism Quebec) (Gill 1997). As with most RRSs, hotel and service listings were provided free of charge, the tourism supplier then paid a commission of 10 per cent to the reservation service as bookings were generated. A feasibility study forecast that RQ would increase the volumes of total reservations to Quebec by 1 per cent – translating into an additional $26 million for the industry (Gill 1997). Despite these high expectations, RQ failed after only 15 months of service. While several of the hoteliers (especially small and medium-sized operators) surveyed in both 1992 and 1994–5 felt that the general concept behind RQ was a good one, they also felt that the tourist industry had not been sufficiently involved in the planning and development process. In light of its poor experience with RQ, Tourism Quebec has since retreated from RRSs and is again focusing on its 1–800 information service.

Alliances and networking

One way in which independent operators can attempt to overcome the problem of CRS-access is by joining a voluntary reservation consortia which provides clients with a reservation system similar to those used by large chains. The companies that run these networks have become major players in the global hotel industry, with several of the largest operators often being based outside the US (Table 10.5). It is important to note, however, that the membership costs of such systems remain prohibitive to many small operators (see Chapter 7 in this book). Hoteliers in Montreal who do not have CRS-access are forced to turn to new distribution outlets, such as the Internet, or to create their own networks/alliances that provide establishments with the potential to gain improved access to the global travel distribution system. While the use of the Internet is growing, we found little evidence of groups of firms attempting their own independent pooling of resources in order to gain access to CRS.

Table 10.5 The world's ten largest voluntary reservation consortia, 1995

Company	Rooms	Hotels	HQ location
Utell International	1,580,000	6,500	UK
JAL World Hotels	152,809	378	Japan
Lexington Services Corp.	142,000	1,200	USA
Supranational Hotels	117,000	650	UK
Anasazi Travel Resources Inc.	106,400	514	USA
VIP International Corp.	105,243	971	Canada
Leading Hotels of the World	87,000	300	USA
Logis de France	71,960	4,050	France
SRS Hotels Steigenberger	65,000	330	Germany

Source: *Hotels* (1996: 72)

In an attempt to improve access to the travel distribution system, some small bed and breakfasts and boutique hotels were beginning to develop marketing alliances, usually in the area of shared brochures. In these cases the key factors underlying alliance formation appear to be location, type of hotel, and the long-term development of inter-firm relations. In particular, the development of trust and reciprocity between potential partners is vital, every bit as vital as it is in more 'innovative' segments of the economy (Hansen 1992).

Large and medium-sized hotels also have an effective lobbying alliance in the shape of the HAGM. This grouping paid $40 million into municipal taxes during the 1990–91 financial year (source: HAGM interviews) and as such wields significant power at the municipal level. The HAGM has been particularly vociferous on issues such as municipal taxes and the federally imposed goods and services tax. An association of small hotels had also recently been formed in Montreal to provide a collective lobbying voice against unpopular municipal and provincial policies. As a grouping these particular hotels represent more than 300 beds, putting them on the same footing as one of the larger hotels.

Other global marketing alliances are affecting the Montreal hotel industry. Managers of larger enterprises stated on several occasions that frequent flyer (FF) programmes had revolutionized the industry through its encourage-ment of repeat visitation and brand loyalty. While every international chain affiliate was linked to at least one FF programme, no small independent operators had joined. This reflects the fact that hotels require significant economies of scale to be able to afford the costs involved in joining these global partnerships. While many hotels also had 'frequent stay' programmes in place, their use was closely tied to the size of the operation (see Toh *et al.* 1991; Dev and Klein 1993). Only 9 per cent of the small hotels had any programme that recognized repeat guests, compared to 40 per cent of their medium-sized counterparts. Nearly four-fifths of large hotels were operating some form of programme.

Hotels, both large and small, have also been at the centre of increasing network formation between different segments of the urban tourism product. Because of their central location, cultural (and often historic) significance, and pivotal tourist role, hotels in Montreal often form close partnerships with nearby cultural attractions and with the city's myriad special events. One downtown chain is associated very closely with the city's world famous jazz festival, while others have created close links with the theatre and film sectors. A vital component in the success of these networks is spatial proximity, with some hotels actually becoming integral components of the cultural events they are linked to.

Conclusions

An evolving competitive environment driven by market saturation, recessionary pressures, and shifting consumer demand led Montreal hotels to introduce a range of organizational changes in the early to mid 1990s. Hotels have attempted to segment the market place in innovative ways, and are using information technologies to improve service levels, strengthen data-gathering capabilities, and reach new markets. At the time of the survey, management were also beginning to question traditional approaches to labour use in the industry, with innovative companies attempting to reduce turnover rates and improve staff performance through incentive-based reward schemes and other compensatory mechanisms. The formation of alliances and networks at a variety of scales – from the global (CRS), to the local (hotel-attraction links) – was also a key response to the evolving competitive environment.

While there is little doubt that shifting demand and a variety of macro-economic factors had created a somewhat different competitive environment, the responses adopted by the Montreal hotel industry do not represent a clear break from past organizational structures. Labour processes have not changed dramatically and there seems little doubt that female employees working for relatively low wages will continue to dominate the industry. Temporary and part-time labour use is not new to the sector, and is in fact declining in significance. Low-skill, semi-skilled and middle-management positions are being reduced gradually as larger companies attempt to flatten their organizational structures. The skills requirements of those workers remaining are in many cases rising because, as Schlesinger and Heskett (1991: 47) note: 'the more that technology becomes a standard part of delivering services, the more important personal interactions are in satisfying customers and in differentiating competitors'.

It is crucial that ongoing research be conducted on the human resource issues that stem from these findings, and that recommendations be made as to how both the private and public sector can cope with the challenges that lie ahead. While it is not generally a problem finding a sufficient number of people to fill low skill positions (e.g. room cleaners), certain areas of hotel operation appear to be facing shortages of skilled staff. Some of our respondents felt that labour shortages had been exacerbated by the failure of industry and public-sector training programmes to adequately respond to changes in technology and consumer needs. The Federal Government clearly recognizes this problem, and is beginning to develop strategies to address it (see Employment and Immigration Canada 1994).

We saw little evidence to suggest that levels of vertical disintegration will increase significantly in the Montreal hotel industry. There is also little evidence in the Montreal case to support the notion, raised in much of the flexibility literature (see Loveman and Sengenberger 1990), that smaller, networked companies will be better placed to respond to the shifting competitive environment than their larger counterparts. Our findings reveal that large multi-national and national chains are best able to afford access to the global distribution systems that play such an important role in a firm's ability to reach the market. Certainly it is larger chains that have the greatest chance to move into new markets in an attempt to counteract the intense competition found in established destinations. For example, many US and European chains have been at the forefront of investment into Russia where, at least initially, a lack of quality, domestic competition and a booming, business travel market allowed high profit margins, and some respite from the often low returns of the West (Milne and McMillan 1997a). While small specialized firms can also be expected to survive and perhaps even thrive in some niche markets, it is medium-sized, independent operators who face the bleakest prospects.

Attempts by smaller operators to form alliances and networks may assist them in their efforts to remain competitive. It is clear that, although these formations are very fragile and require considerable fostering, the area of net-working and alliance formation will remain vital to the future success of the hotel industry in Montreal and elsewhere. While we have quite a reasonably good sense of the evolving international structure of the hotel industry and the development of global alliances (see Gannon and Johnson 1995; Littlejohn and Beattie 1992), we still know relatively little about the factors that allow firms operating within spatially proximate areas to create successful partnerships. There clearly needs to be more attention paid by tourism researchers and economic geographers to this area of research.

While many of the processes identified above are clearly unique to this particular consumer service, and the setting within which it operates, they nevertheless call into question the broader sectoral applicability of current theories of economic change that have gained so much popularity among economic geographers in recent years (such as regulation theory and flexible specialization). We certainly, for example, cannot as yet speak of a radical shift from 'Fordist' to 'post-Fordist' organizational structures within the Montreal hotel industry. Core patterns of labour-use, levels of vertical disintegration, and the relative competitive and market-place positioning of small and large firms have all remained fairly constant. But we have also witnessed considerable change in other areas – much of it influenced by the introduction of new technologies.

It is this complex interaction of continuity and change that we must seek to address as economic geographers. This chapter has shown that by extending our focus beyond those sectors that have traditionally captured the attention of economic geographers, and by developing a broader and more detailed understanding of processes of economic change occurring *within* consumer service firms such as hotels, we can add a number of potentially important strands to our understanding of current processes of economic restructuring occurring in advanced economies. It is only by gathering more of this evidence from areas like the tourist industry that we can begin to answer Marshall and Wood's (1992) call for the development of a more 'services informed' approach to understanding current processes of urban economic and social change in Canada and elsewhere.

Acknowledgments

The research presented here was supported by grants from the Social Sciences and Humanities Research Council of Canada, the Quebec FCAR, and the McGill Faculty of Graduate Studies. We would like to thank the various managers and industry association representatives who participated in the interviews. We would also like to thank Kara Gill and Stephen Tufts of the McGill Tourism Research Group for their valuable research assistance.

Part D

GLOBAL–LOCAL NEXUS: PLACE COMMODIFICATION, ENTREPRENEURSHIP, AND LABOR

11

THE INSTITUTIONAL SETTING – TOURISM AND THE STATE

C. Michael Hall

The concept of the 'state' is one of the key elements of analysis in human geography (R. Johnston 1982). Its character, role, function and influence have left an indelible mark on the location and nature of economic activity, welfare, personal geographies and the cultural landscape. However, despite the critical role of the state on individual and collective actions, the concept has received little substantive attention in the tourism literature (Hall and Jenkins 1995). Indeed, as Richter (1983) described elsewhere with respect to tourism and politics, research on the role of the state in tourism has been sadly neglected.

The nature of the state

The state is the formal mechanism through which political power is exercised and which commands obedience (Curtis 1962). The state lies at the inter-section of sovereignty and territoriality. According to P. Taylor (1994: 151), 'the power of the modern state is based to a large degree upon the fusing of the idea of state with that of nation to produce the nation-state'. Giddens (1985), in relating the exercise of power by the state within a given territory, described the state as a 'power container'. The state can be conceptualised as a set of officials with their own preferences and capacities to effect public policy, or, in more structural terms, as a relatively permanent set of political institutions operating in relation to civil society (Nordlinger 1981). Therefore, the state is essentially the entire apparatus of formal roles and institutions which exercise authority over populations in a given territory. It includes elected politicians, the various arms of the bureaucracy, public/civil servants, and a plethora of rules, regulations, laws, conventions and policies. The functions of the state will affect tourism to different degrees. However, the degree to which individual functions are related to particular tourism policies, decisions and developments will depend on the specific objectives of institutions, interest groups and significant individuals relative to the policy process (C.M. Hall 1994). The state performs many functions:

- as developer and producer;
- as protector and upholder;
- as regulator;
- as arbitrator and distributor;
- as organiser (Davis *et al.* 1993).

Each of these functions affects various aspects of tourism including development, marketing, policy, promotion, planning and regulation. Two important themes in tourism research, which implicitly address the issue of the regulatory role of the state in tourism, are those of the appropriate role of public-sector tourism agencies, and the search for sustainability at different policy and planning scales (macro, meso and micro) (Hall and Jenkins 1995). However, the state often fails to act in a coordinated way, with some decisions and actions being confusing to many observers. For example, while one arm of government may be actively promoting environmental conservation in a given location, another may be encouraging large-scale tourism development in order to attract employment and investment. Furthermore, there are limits to effective state intervention in economy and society, particularly as the effects of globalisation are felt by individual cities and regions at the sub-national level, also referred to as the 'local state'. Different policies in different policy arenas are often in conflict, and different levels of state action, particularly in a federal system, may have conflicting goals, objectives, outcomes and impacts (Jenkins and Hall 1997a), therefore raising substantial questions about the possibility of coordinating an industry as diffuse as tourism.

Effective coordination of tourism policies between different levels of the state is unlikely. For example, within the Australian federal system, the emphasis on state rights has meant that even when state and federal governments are all of the same political persuasion, unanimous agreement on policy settings is still extremely difficult to achieve (C.M. Hall 1995). As Craik recognised:

> Both the domestic and foreign airlines have been partially de-regulated, employment conditions are being liberalised, foreign investment restrictions have been freed up, and more money has been put into tourism marketing and promotion. But such initiatives lack coordination and coherences as well as considered objectives and outcomes.
>
> (Craik 1991: 235)

Coordination

Coordination is a very difficult task because the problems involved are not only administrative, but political (C.M. Hall 1995). As Hogwood and Gunn (1984: 205–6) recognised, 'coordination is not, of course, simply a matter of communicating information or setting up suitable administrative structures, but involves the exercise of power'.

Coordination is about power and politics. Numerous individuals and organisations (public and private) are seeking control or influence over particular tourism policies or issues. These stakeholders have a view on what planning and development perspectives or approaches are best and will want to ensure access to the resources they require for their preferences or approaches – a situation in which, if they are successful, will increase their control over decisions and actions and thereby strengthen their power base. Unfortunately, however, very little is known about the power base for decisions and actions relating to tourism development (Hall and Jenkins 1995). Therefore, very little is known about how actual decisions are made.

Tourism policymaking and planning have been described by Jenkins and Hall (1997a) as constituting a 'wicked task' where numerous values and interests compete over scarce resources.

> Wicked problems have no definitive formulation and hence no agreed-upon criteria to tell when a solution has been found; the choice of a definition of a problem, in fact, typically determines its 'solution'. . . . Since wicked problems are subject to innumerable political definitions, there are no ultimate tests to measure the efficacy of their solutions.
>
> (Harmon and Mayer 1986: 9–10)

Despite this situation, much tourism research neglects the institutional arrangements for tourism. Perhaps this situation can be explained by tourism's relatively recent rise to economic, social and political prominence. Whatever the case, an understanding of the role that the institutions of the state play in tourism is vital in understanding the effects of tourism policy on the patterns and processes of tourism development.

Tourism and the institutions of the state

Understanding tourism policy and planning processes, outcomes and impacts requires reference to the institutional arrangements for tourism policy (Figure 11.1). Such arrangements vary significantly between countries and between policy arenas within an individual country. These differences affect the content and nature of overt and covert political debate and argument, the strategies individuals and groups employ in attempting to influence policy (spatially, contextually and over time), and the weight that policymakers ascribe to particular social and economic interests (Brooks 1993). The institutional framework mediates conflict by providing a set of rules and procedures that regulates how and where demands on tourism policy can be made, who has the authority to take certain decisions and actions, and, ultimately, how policies and plans are implemented. It also expresses conflict,

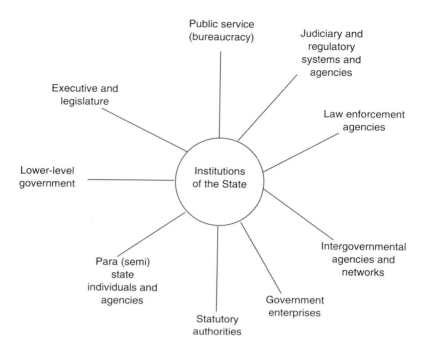

Public service (bureaucracy)

Judiciary and regulatory systems and agencies

Executive and legislature

Law enforcement agencies

Lower-level government

Institutions of the State

Para (semi) state individuals and agencies

Intergovernmental agencies and networks

Government enterprises

Statutory authorities

Executive and legislature: e.g. systems of government, heads of state, government and opposition, minister responsible for tourism.

Public service (bureaucracy): e.g. government departments (and their staff), departments of tourism, tourism bureaucrats.

Judiciary and regulatory systems and agencies: courts of law.

Law enforcement agencies: armed forces, police, customs.

Intergovernmental agencies and networks: committees, councils, conferences, networks and partnership (formal and informal), ministerial committees on tourism.

Government enterprises: trading banks, essential services (e.g. communications and transport), statutory travel and tourism promotion organisations.

Statutory authorities: central banks; educational institutions (schools and higher education), regional development authorities.

Para (semi) state individuals and agencies: media, interest groups, trade unions, peak industry bodies, chambers of commerce, regional tourism industry associations.

Lower-level government: state/provincial and local/regional governments.

Figure 11.1 Schematic structure of institutions of the state with reference to tourism
Source: Based on Hall and Jenkins (1995: 20)

in that institutions and relationships comprising the state system reflect and adapt to the broader pattern of social and economic forces (Brooks 1993).

The setting of the rules by which policy debate is carried out, in other words the 'rules of the game', is a major function of the state as the rules serve to legitimise the outcome of debate (Hall and Jenkins 1995). The 'rules of the game' are 'a set of predominant values, beliefs, rituals and institutional procedures that operate systematically and consistently to the benefit of certain persons and groups at the expense of others'; those who benefit 'are placed in a preferred position to defend and promote their vested interests' (Bachrach and Baratz 1970: 43–44). The rules of the game affect the behaviour of groups.

As intervening factors between the individual political actor and the generalised norms of the political system, these rules have certain distinctive attributes and comprise an important part of the policymaker's environment. '[These rules are linked to] the values, myths and beliefs that tie a people together in a community and the overall pattern of power and authority relationships in the society as a whole' (Anderson 1984: 278).

Therefore, instutionalized rules condition the cognitive and normative understandings of different political actors, thereby facilitating some strategies while constraining others. Institutionalised rules refer to accepted ways of viewing how society works and consequent prescriptions for attaining objectives. For example as Hall and Jenkins (1995) note, a parliamentary or congressional committee may be established to enquire into certain issues surrounding the tourist industry, but the terms of reference which are set by the committee will determine the scope of their policy discussions and therefore the outcomes. As Schattschneider (1960: 71) has written: 'All forms of political organisation have a bias in favour of the exploitation of some kinds of conflict, and the suppression of others, because organisation is the mobilisation of bias. Some issues are organised into politics while some others are organised out.'

The institutions of modern states take on a number of forms, roles and responsibilities. The extent of the state's role in tourism varies according to the conditions and circumstances peculiar to each country (politico-economic-constitutional system, socio-economic development, degree of tourism development) (IUOTO 1974). Indeed, IUOTO argued that tourism was such a key sector that in order to foster and develop tourism:

> . . . on a scale proportionate to its national importance and to mobilise all available resources to that end, it is necessary to centralise the policy-making powers in the hands of the state so that it can take appropriate measures for creating a suitable framework for the promotion and development of tourism by the various sectors concerned.
>
> (IUOTO 1974: 71)

Nevertheless, the optimism of the above comment fails to recognise that the state is also an arena of conflicting values and interests. Further complicating the role of the state in tourism is the number of policy areas (e.g. transport, aviation, immigration, environment, industrial relations, foreign affairs and trade) that are related to tourism, but often do not have tourism-specific interests. Indeed, the implications of the difficulty in actually defining what tourism is have not been adequately thought through in terms of the inability of the state to intervene in the tourism policy arena (C.M. Hall 1994; Hall and Jenkins 1995; see chapters in Part A in this book). The concept of 'partial industrialisation' is one attempt to describe such difficulties and the consequent problems of coordination, management and strategic development. According to Leiper (1989: 25), partial industrialisation refers to the condition in which only certain organisations providing goods and services directly to tourists are in the tourist industry. The proportion of (a) goods and services stemming from that industry to (b) total goods and services used by tourists can be termed 'the index of industrialisation,' theoretically ranging from 100 per cent (wholly industrialised) to zero (tourists present and spending money, but no tourist industry).

One of the major consequences of the partial industrialisation of tourism is its significance for tourism development, marketing and coordination. Although we can recognise that many segments of the economy benefit from tourism, it is only those organisations with a direct relationship to tourists that become actively involved in fostering coordinated approaches to tourism development or promotion. However, there are many other organisations such as food suppliers, petrol stations, and retailers, sometimes described as 'allied industries', which also benefit from tourists but which are not readily identified as part of the tourist industry. According to Leiper (1990b: 163–64) such a situation provides an underlying reason 'why there are ongoing calls for more governmental involvement despite growth in tourism. . . . In principle, what governments are asked to do is make up the non-industrialised parts of supply and its management.' However, this then raises a further question, 'why should governments help one of the biggest industries . . . when there are many new and small industries that may have better claims and when there are pressing social items in the agenda?' (Leiper 1990b: 164). Furthermore, the partially industrialised nature of the tourist industry, if indeed it really is an industry, also indicates the difficulties of peak industry associations being representative of the full range of businesses which depend on tourists and, therefore, in articulating their claims on government. However, even if claims for tourism could be clearly articulated, the fragmented nature of tourism and the diffuse manner in which it affects a wide range of policy issues indicate the difficulties for consistent and coordinated state action.

But is it possible for the state to act consistently anyway? Marsh (1983) clearly dismissed the view that the state is a consistent, coherent, cohesive institution, either with a common self-interest, or with a capability of deciding what is best in the national interest: 'any empirical observation . . . shows that [states] are not undifferentiated wholes, there are more or less important divisions within the executive, and between the executive and the bureaucracy' (Marsh 1983: 12). As Williams and Shaw (1988: 230) observed with respect to tourism, 'the aims of the local state may diverge from those of the central state'. This is particularly apparent in federal systems, such as those of Australia, Germany, India and the United States, where there exist three (and arguably four) levels of the state: national, state (regional), and local. Put simply:

> In a federal system, which disperses power among different levels of government, some groups may have more influence if policy is made at the national level; other groups may benefit more if policy is made at the state or provincial level . . . institutional structures, arrangements, and procedures can have a significant impact on public policy and should not be ignored in policy analysis. Neither should analysis of them, without concern for the dynamic aspects of politics, be considered adequate.
>
> (Anderson 1984: 18)

The actions of the state with respect to tourism are therefore forged and shaped principally within a complex arrangement of political and public institutions, and with varying influence from interests in the private sector. However, as noted above, the state should not be seen as a unitary structure. The sources of power in tourism policy, planning and promotion affect the location, structure and behaviour of agencies responsible for tourism policy formulation and implementation. The diversity, complexity and changing nature of the tourist industry, and the community and state generally, results in unstable power relationships. For example, in Australia, as with other Western countries such as Canada, New Zealand and the United States, there have been substantial tourism policy and administrative shifts at all levels of government. In 1974 tourism at the federal level was part of the Department of Tourism and Recreation; in 1985, tourism was in the Department of Sport, Recreation and Tourism; in the late 1980s, tourism became part of the Department of the Arts, Sport, the Environment, Tourism and Territories; and in 1991 tourism finally became a separate federal government department. In 1996, with the Liberal-National Party coalition coming to power for the first time in thirteen years, the Department of Tourism was abolished and amalgamated with the Industry and Science portfolio (Jenkins and Hall 1997a). As Mercer (1979) observed:

> The setting up of entirely new government departments, advisory bodies or sections within the existing administration is a well-established strategy on the part of governments for demonstrating

loudly and clearly that 'something positive is being done' with respect to a given problem. Moreover, because public service bureaucracies are inherently conservative in terms of their approach to problem delineation and favoured mode of functioning . . . administrative restructuring, together with the associated legislation, is almost always a significant indicator of public pressure for action and change.

(Mercer 1979: 107)

The Australian experience has been closely replicated in New Zealand. Two government organisations are directly involved in tourism in New Zealand: the New Zealand Tourism Board (NZTB) which is responsible for international marketing and promotion, and the Tourism Group of the Ministry of Commerce which is responsible for policy advice to government. The split between promotion and policy occurred in 1991 when the New Zealand Tourism Department (NZTD), which was itself formed in July 1990 from the restructured New Zealand Tourist and Publicity Department, was abolished (Pearce 1992). The policy function was initially given Ministry status, but this was later downgraded to divisional status in the Commerce Ministry. As Chamberlain (1992: 91) observed: 'reforming, renaming, restructuring, re-logoing and relaunching tourism is an ingrained tradition' in New Zealand. In Canada, Tourism Canada has been established in order to give greater promotional and corporate emphasis to government's tourism activities, while in the United States the national tourism administration was abolished with state governments and industry being encouraged to play a more active role in tourism promotion in partnership with the federal government.

Changes in the place of tourism in the machinery of government and in the structure, aims and objectives of such organisations is a familiar occurrence at all levels of government. Governments have struggled to resolve their roles in tourism planning, development, marketing and promotion, and in dealing with the often vociferous claims of the tourist industry on a wide range of public policy issues. This situation is reflected in the way in which governments frequently alter public sector arrangements for tourism planning, development, marketing and promotion. Such a situation has been well described by Hogwood and Peters:

Public ends are ambiguous and shifting, the means are unreliable and controversial; decision rules may vary from one situation to another. Instead of rational processes for the making and implementation of public policies we may expect to find irrational events in the public sector: solutions looking for problems, participants searching for choice opportunities and outcomes which seek some relation to organisational [and more generally political] goals.

(Hogwood and Peters 1985, in Lane 1993: 4)

As noted above, changes to the institutional arrangements for tourism in which demands for less state intervention have been met with government attention being focused on tourism marketing and promotion, are seen in much of the Western world. Tourism planning and development functions have tended to be abolished at the central level and are increasingly being assumed at the sub-national level. Reasons for such a shift, however, probably lie not so much with the failure to recognise the economic and political significance of tourism but with broader changes in ideology and philosophy about the appropriate size and role of the state in the broader political environment within which tourism operates.

Within Western society considerable debate has emerged in the past two decades over the appropriate role of the state in society. Throughout most of the 1980s, the rise of 'Thatcherism' in the United Kingdom and 'Reaganism' in the United States saw a period of retreat by central governments from active government intervention. At the national level, policies of deregulation, privatisation, the elimination of tax incentives, and a move away from discretionary forms of macro-economic intervention, were the hallmarks of a push towards 'smaller' government and the supposed withdrawal of government from the economy and the market-place (Hall and Jenkins 1995; Jenkins and Hall 1997b).

Tourism is not immune from such changes in political philosophy. Tourism is subject to direct and indirect government intervention often because of its employment and income-producing possibilities, and therefore its potential to diversify and contribute to regional economies. Given calls for smaller government in Western society in recent years, there have been increasing demands from conservative national governments and economic rationalists for greater industry self-sufficiency in tourism marketing and promotion, often through the privatisation or corporatisation of tourism agencies or boards (Jeffries 1989). For example, Tourism Canada has been reorganised along corporate lines from its previous public-service structure, with substantial emphasis being placed on industry involvement in the new organisation not only in terms of industry funding for marketing and promotion but also strategic direction (Smith and Meis 1997). Similarly, in Australia, state governments have developed event-bidding corporations as private companies which, although established by state governments, have only a narrow economic mandate with respect to the attraction of major events (e.g. Konrads 1993). The implications of such an approach for the tourist industry are substantial. As H. Hughes (1984: 14) noted with respect to Thatcher's Britain of the early 1980s, 'The advocates of a free enterprise economy would look to consumer freedom of choice and not to governments to promote firms; the consumer ought to be sovereign in decisions relating to the allocation of the nation's resources.'

The rise of the new right 'enterprise' philosophies of Thatcher and Reagan in the 1980s and the emphasis on 'smaller' government went hand-in-hand with the emergence of globalisation as a buzzword in economic and political circles and a rediscovery of places and regions as a unit of economic development. Such a relationship was no accident. Impetus for freer intra-national trade and a lessening of government intervention was acted out at the international sphere as a desire from some governments and economists for a lowering of tariff barriers and other forms of protection in order to encourage more efficent and freer international trade.

According to Kotler *et al.* (1993) we are living in a time of 'place wars' in which localities are competing for their economic survival with other places and regions not only in their own country but throughout the world (see plates 11.1–11.3)

> All places are in trouble now, or will be in the near future. The globalisation of the world's economy and the accelerating pace of technological changes are two forces that require all places to learn how to compete. Places must learn how to think more like businesses, developing products, markets, and customers.
>
> (Kotler *et al.* 1993: 346)

In the globalised economy, competition is therefore essential not only within the nation but also outside it. In this environment, Giddens's (1985) notion of the nation state as a container no longer holds water. The better analogy may be that of the nation state as a sieve. In this environment, therefore, regions and places at the local state level become the key unit of competition not only for investment and employment but, increasingly, for tourists.

Tourism and the local state

The local state is 'the set of institutions charged with the maintenance and protection of social relations at the sub-national level' (Johnston *et al.* 1986: 263). The sub-national level, of course, should not be seen as a singular level of territoriality. Government may have several levels at the sub-national level including provinces or states, regional government and local government. However, within the globalised economy, the local state, in its various forms, is almost universally seen as having increased importance. For example, new centres and peripheries and also new territorial hierarchies, have been created by geographical transformations. These transformations are being brought about through the international restructuring of capitalist economies and the consequent changes to the nature and role of cities and regions, as they seek to attract ever more mobile investors. New relational contexts and configurations have resulted; beyond this, there is the overarching global context and 'regional differentiation becomes increasingly organised at the international rather than national level; sub-national regions increasingly give way to regions of the global economy' (S. Smith 1988 in Robins 1991: 24).

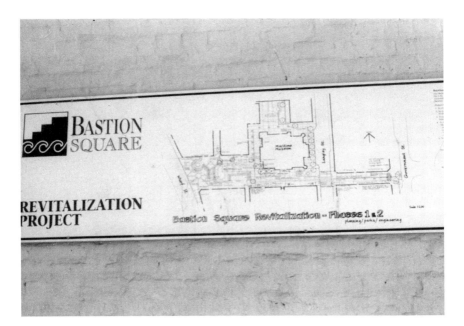

11.1 Revitalisation project in Victoria, British Columbia: Waterfront regeneration for the visitor dollar

Source: Michael Hall

11.2 Flattening the waterfront: interpretive panels in Hobart, Tasmania, Australia. The panels provide a history of commerce and the enterprise economy of the Hobart waterfront. However, the long history of labour disputes and the poor conditions of the workers are virtually ignored

Source: Michael Hall

11.3 Relict heritage: waterfront crane at the London Dockland redevelopment. The crane is meant to remind us of the industrial heritage of the site. Unfortunately, the crane is now almost completely out of the heritage context and is left as an architectural gesture to provide what little is left of 'local flavour'
Source: Michael Hall

Tourism is intimately connected to the place-marketing process because of the way in which it is often used as a focus for urban redevelopment, revitalisation and promotion strategies. Tourism, as one of the major components of the expanding service economy, is often regarded by government and some academic commentators as a major mechanism for attracting investment, creating employment and promoting regional economic growth (e.g. see Kotler *et al*. 1993).

The competitive ethos of the market place has therefore became translated into a burgeoning 'place market' (Sadler 1993). 'The primary goal of the place marketer is to construct a new image of the place to replace either vague or negative images previously held by current or potential residents, investors and visitors' (Holcomb 1993: 133), in order to compete effectively with other places within the constraints of a global economy for a share of mobile international capital (Harvey 1989b) (Plates 11.1–11.3). 'This marketing operation involved the construction or selective tailoring of particular images of place, which enmeshed with the dynamics of the global economy and legitimised particular conceptions of what were "appropriate" state policy responses' (Sadler 1993: 175). The new rhetoric of 'the local' therefore has to be seen as deeply embedded in processes of global accumulation, in a 'fragmented mosaic of uneven development in which competitive places try to secure a lucrative development niche' (Swyngedouw 1989: 31).

Ideas of place marketing are closely related to notions of flexible production and consumption inherent in ideas of post-Fordism. Within the tourism context, these ideas have received considerable attention in the work of Poon (1993) in the Caribbean and, to a lesser extent, in texts such as that of Shaw and Williams (1994). The focus on flexibility has considerable significance to the internal and external operations of tourism organisations. For example, internal labour-market organisation in tourism and hospitality is often highly flexible because of seasonal and variable patterns of demand (Shaw and Williams 1994). Externally, the concept of flexible production is tied in the production of places as commodities to be promoted and sold to consumers. 'Place marketers do not see their task as purely promoting and advertising, but also as adapting the "product" (that is, the place) to be more desirable to the "market"' (Holcomb 1993: 134). For Kotler et al. (1993), flexibility is inherent in the notion of place marketing:

A marketing approach to place development is the overriding response places need to compete effectively in our new economy. Places must produce products and services that current and prospective customers want or need. Places must sell products and services internally and externally, nationally and internationally. Place marketing is a continual activity that must be adjusted to meet changing economic conditions and new opportunities.

(Kotler et al. 1993: 345)

Similarly, Fretter (1993: 165), a local government employee in Britain, argued that place marketing 'should aim to ensure that urban activities and facilities are related as closely as possible to the demands and desires of targeted customers and clients [which] requires a more flexible approach to development plans'. However, although the notion of flexible economic response, whether it be by firms or places, is one of the dominant notions of the enterprise culture being promoted by governments in the Western worlds (see Corner and Harvey 1991), our understanding of contemporary economic processes is weakened by the lack of attention paid to economic activities in the service sector and the narrow range of regional settings in which studies have been conducted. Indeed, our understanding of flexibility in advanced capitalism needs to be considered within the context of global accumulation and a political economy of tourism, rather than within a more narrow, empiricist, economic and marketing tradition (C.M. Hall 1997).

The notion of rapidly circulating international capital within the global economy is also implicit in the work of Kotler *et al.* (1993) which comes from within the mainstream, empiricist, marketing tradition. According to Kotler *et al.* (1993: 18), a central proposition of their book *Marketing Places: Attracting Investment, Industry, and Tourism to Cities, States, and Nations* was that the 'marketplace shifts and changes occur far faster than a community's capacity to react and respond. Buyers of the goods and services that a place can offer (i.e. business firms, tourists, investors, among others) have a decided advantage over place sellers (i.e. local communities, regions, and other places that seek economic growth)'.

Kotler *et al.* (1993: 18) refer to the need for places to adopt a process of 'strategic place marketing' for urban and regional revitalisation in order to design a community 'to satisfy the needs of its key constituencies' by embracing four core activities:

- Designing the right mix of community features and services.
- Setting attractive incentives for the current and potential buyers and users of its goods and services.
- Delivering a place's products and services in an efficient, accessible way.
- Promoting the place's values and image so that the potential users are fully aware of the place's distinctive advantages.

All of this has substantial implications for the actions and institutional arrangements of the local state and, indeed, the nation state. In commodifying place as a product that can be revitalised, advertised and marketed, places are presented not so much 'as foci of attachment and concern, but as bundles of social and economic opportunity competing against one another in the open (and unregulated) market for a share of the capital investment cake (whether this be the investment of enterprises, tourists, local consumers or whatever)' (Philo and Kearns 1993: 18). The competition for investment is not just restricted to the hosting of an Olympic Games or other major events such as the World Cup of soccer or a World's Fair. Place competition is also occurring with respect to encouraging airlines to travel to a city and, where possible, choose a place as a location for a transport hub, and for major tourism and hotel corporations to locate their hotels or other ancillary services in a particular city. In order not only to encourage relocation but also to retain

traditionally been fragmented with a top-down bias whereby governments or agencies, often in partnership with the private sector, promote and develop tourist destinations. With increasing emphasis by government at all levels – on the supremacy of the market, state involvement through public–private partnerships, reduced budget deficits, international trade expansion (especially in growth industries such as tourism), employment growth, and lower levels of direct, central government intervention – local communities (including their locally elected representatives, namely, local government) will be left to ponder what might or should be, and how to get there (Jenkins and Hall 1997b).

The local state, as the immediate and critical receptor for local development plans and development on behalf of its constituents, will be required to conduct its functions in increasingly dynamic environments, perhaps with less direct assistance from higher levels of government. But at the same time it is being strongly urged by central government and other interests to take up tourism opportunities, especially opportunities associated with large-scale developments or events. However, the decision-making process for local communities can be confusing. It is therefore difficult to see how tourism-specific policies, and especially those of small, under-resourced local governments:

> . . . can counter processes in which place prosperity is largely determined by exogenous events like changing lifestyle preferences, infrastructure investment, technological innovation, currency movements (a major factor affecting farm profitability and the flow of overseas tourists), interest rates, taxation, the length of the working week, labour costs, and so on.
>
> (Sorenson 1990: 59)

While it may sound clichéd, tourism must be regarded as a means to an end. The most appropriate policy and institutional response to the global environment will be one that sees tourism as part of an integrated, inclusive, development strategy, rather than as a single end in itself.

Conclusions

The search for comprehensive integration between the institutional arrangements for tourism and tourism policy (e.g. in marketing and promotion; administration and management of impacts; planning and development) has met with little success. The continuing fragmentation of responsibility for industry policy and planning, and the parallel fragmentation of the industry, ensure the status quo of overlapping policy goals and instruments. There is a clear need for coordination in all areas of the tourist industry and not just in marketing and promotion. The nature of state involvement in tourism has significant implications for the sustainability of tourism development. As governments at all levels struggle to clearly identify their roles, responsibilities, purpose and objectives in the global environment, the tourism policy process and the design of appropriate institutional arrangements becomes more tangled. State responses are less likely to be medium-to-long term, and are less likely to encourage well focused policy debate and argument. Interest groups also find it difficult to determine to whom their energies should be best directed. Although a separate tourism agency or department may be set up as the focal point for one or more activities (e.g. tourism research, planning, development or promotion), that point will more than likely involve a number of competing interest groups representing diverse and contradictory demands (Brooks 1993). This may help the agency to create a public perception that it tries to isolate itself from specific interests. On the other hand, it may simply ensure that whatever the agency does it will be criticised by somebody – it will be 'damned if it does' and 'damned if it does not' (Jenkins and Hall 1997a).

Of course it is extremely unlikely that tourism business interests will be formally integrated into the state. Therefore, the most likely medium-term scenario, particularly if politics becomes more influenced by economic rationalism as we progress through the 1990s, are forms of market-inspired, company-led adjustment strategies, and the continued development of public–private partnerships in selected areas of tourism marketing, promotion and development (Jenkins and Hall 1997a and 1997b). Within this context Poon's (1993) enthusiasm for the development of new 'flexible' forms of tourism may be misplaced. The emergence of more flexible forms of tourism do not necessarily signal more appropriate forms of tourism in terms of their social, environmental or even economic well-being and equity for host communities. The changing nature of the state, with the development of new forms of tourism production and consumption, need to be seen within broader concerns about the nature of contemporary capitalism and the manner in which values, interests and power shape tourism. New structures for tourism may well just mean new structures of power and domination. Amid these developments tourism planning and policy will become an increasingly complex challenge, but one which will become of even more importance for sustainable, socially-responsive and inclusive tourism development strategies. At the local level, a number of questions have to be addressed, including:

- How can tourism development be undertaken inclusively with community acceptance?
- How can local communities come to terms with the wide range of interests involved in tourism planning and development (e.g. those with a financial interest, those representing concerns for environmental impacts, and those with social concerns)?

- How can local communities cope when local interests might not be representative of the national interest (i.e. local state versus nation state), especially when communities have to deal with global and national developments in economic/financial markets, as well as the demands, 'expertise' and resources of bureaucracies and multinational corporations?

(Jenkins and Hall 1997a.)

This is by no means an exhaustive list, but the message is clear. If tourism is to improve the economic and social welfare of local communities, and maintain or enhance the quality of the physical environment of any region, then there is a need for much research on the institutional and policy arrangements for tourism. Such research and associated information-dissemination need to be closely tied to local needs rather than to those of the narrow interests of growth coalitions which the local state has, for the past decade or so, been trying to attract.

Acknowledgments

The author gratefully acknowledges the assistance of John Jenkins, Dave Crag and Kirsten Short in the development of material and ideas in this chapter.

12

TOURISM AND ECONOMIC DEVELOPMENT POLICY IN US URBAN AREAS

Robert A. Beauregard

Introduction

In the 1950s few city governments in the United States viewed tourism as contributing to local economic growth. Consequently, tourism did not figure in their economic development policies, rudimentary as they were. The older cities of the north-east and mid-west focused their energies on urban renewal, retention of their manufacturing base, and improving their infrastructure. Faced with growing populations and thriving local economies, cities of the south and west worried mainly about whether public services could keep pace. At mid-century, tourism was simply not a factor in government-led, urban economic development (Law 1994).

Today, almost all large cities have elaborate economic development programs and many list tourism as one of their most important economic sectors (Judd 1995; Levere 1996) (see Plates 12.1 and 12.2). Along with business services and technology, tourism is considered a significant contributor to a city's prosperity and image (Pagano and O'M. Bowman 1995: 44–67). With rare exceptions, elected officials believe that tourists – whether they be business travelers, vacationers, convention delegates, or day-trippers attending sports events, shopping, or visiting a museum – can help the local economy.

What has brought about this transformation? Why is this sector of the economy, once insignificant, now a high priority on the agenda of economic development policymakers? The intent of this chapter is to delineate the forces that have transformed tourism from a marginal social activity into a key economic sector. Once confined to a small number of tourist attractions and a few hotel and tour operators, tourism has become a major economic force, demanding the attention of political and economic elites as well as the economic development community that serves them (Fainstein and Stokes 1995; Randall and Warf 1996).

In order to explain this transformation, we need to pay close attention to the ways that cities have restructured over the last half-century. Political leaders and economic development officials in the 1950s responded to the conditions that faced them by applying the prevailing theories of urban growth and decline. This dictated an economic development strategy that marginalized tourism. Instead, urban policy was designed to retain and attract manufacturing, revive central business districts, and make the city more accessible and less costly. The cities of the 1990s pose different challenges. Services ranging from finance to health care are prime economic development targets, and local governments emphasize the city's many opportunities for recreation and consumption rather than its function as a producer of goods (Harvey 1987; Sorkin 1992; Zukin 1995). Combined with new perspectives on urban growth, changing urban conditions have elevated tourism to the top ranks of economic development initiatives.

Between explaining the first phase when tourism was all but invisible and the second where its presence is unavoidable, we need to consider the changes in US society that have set the context for forces operating locally. Let us begin in the 1950s and conclude, some pages later, by reflecting on what these social transformations mean for urban tourism as an economic development initiative.

12.1 Reinventing urban areas for the visitor: the Inner Harbour redevelopment, Baltimore, Maryland
Source: Dimitri Ioannides

12.2 Reinventing urban space for tourist dollars: the National Aquarium, Baltimore, Maryland
Source: Dimitri Ioannides

Tourism's absence

Prior to the 1970s, tourism was all but invisible to most city governments. Primarily the concern of convention and visitor bureaus, local officials considered tourism to be a minor economic activity. Some urban governments subsidized convention centers, many worried about the size of their airports, and a few taxed hotel rooms to fund marketing activities, but for the most part urban policymakers did not believe that it was important to identify and manage a diverse tourism sector and to support it with public infrastructure. Oddly tourism failed the export base test that dominated thinking about urban growth (see also Chapter 2 in this book). Its small and problematic job and tax benefits further sealed its marginal status (Jakle 1985: 245–62).

The low profile of tourism in the early post-Second World War period was partly a function of the economic issues that development officials thought most important: manufacturing, downtown renewal, and infrastructure (Beauregard 1993b; Berson 1982; Eisinger 1988: 85–127; Judd and Swanstrom 1994: 127–50). First, expectations in the late 1940s and early 1950s were that urban manufacturing would rebound from its depression-era performance and build on the contribution it had made to the war effort. Although signs of downsizing were apparent in textile and apparel and in heavy industries such as steel, the impending collapse of these industries was unexpected. Manufacturing was still considered a powerful element in the local economy, particularly due to its reputation as a major exporter of goods and thus importer of capital. Consequently, urban economic development policy focused on creating industrial parks, subsidizing manufacturing firms through tax breaks, and smokestack-chasing.

Second, city governments were determined to re-assert the economic dominance of their downtowns (Teaford 1990). Not only did central business districts (CBDs) constitute a significant portion of the local property tax-base, but they generated regional retail activity and housed the expanding business functions that would soon replace manufacturing as the dominant urban sector. Between 1929, the year of the Stock Market Crash, and the late 1940s, little investment had occurred in CBDs (Beauregard 1989: 3–11). Property values had fallen as buildings deteriorated and businesses began to leave. The downtowns needed to be renewed. Using federal legislation and federal funds, local governments established extensive urban renewal initiatives to boost central-business-district land values and reinvigorate office and retail activity. Urban renewal also included the demolition of residential slums and of blighted properties in mix-use areas. In many instances, local governments financed office buildings, sports stadia, or convention centers in order to create a critical mass of new investments that would anchor private sector development. Urban renewal agencies took charge of CBDs and slum clearance, with economic development agencies left to focus on manufacturing.

Third, the city's infrastructure also had been neglected and suffered from deferred maintenance and obsolescence. Older cities needed to upgrade their sewage and water systems, build modern airports, construct (with federal and state monies) highways to accommodate rising automobile usage, and attend to their waterfronts. New and renewed infrastructure would retain businesses and attract investors. A redeveloped downtown would expand office activities and retain retailers, consequently helping the city hold onto its middle class.

Finally, manufacturing would provide good-paying jobs for the working class and keep them in the city. Except for the few already-established tourist destinations such as New Orleans or San Francisco, tourism was not a target of public economic development initiatives.

Dominating the prevailing theory of urban economic development was export base theory (Thompson 1968: 11–60). Simply, it divides the local economy into: basic industries that export goods and thereby bring capital into the community; and non-basic industries that serve the basic industries, their workers and the workers' families. Of major importance is the subsequent circulation of new money through the local economy, a phenomenon captured in the notion of the multiplier; simply, the number of times an 'imported' dollar is spent at local businesses. The goal, of course, is a high multiplier and the key to local economic growth is a strong basic sector. In the 1950s and 1960s, manufacturing was considered the quintessential basic industry.

At that time, tourism did not fit the export base perspective (see also Chapter 2 in this book). It produced no goods and, because of its high proportion of low-wage and seasonal jobs, seemed to have minimal impact on the local economy. Services were not considered economically important. Visitors came and spent money, but nothing much else happened, at least policymakers believed. Additionally, the tourist sector had yet to establish a prominent image in cities and its proponents had yet to forge a connection to dominant economic and political elites. Neither defined as a basic industry nor blessed with a noteworthy multiplier, and having little political presence, tourism did not attract the attention of economic development officials.

Although we might make a different argument now and claim a misperception on the part of policymakers and analysts, the fact is that the small size of the urban tourist sector and the undervaluing of its economic importance relegated it to a marginal status. In the early 1950s few cities thought about conventions as substantial draws for visitors, conceived of their historical sites as more than antiquarian oddities, or viewed hotels as serving both tourists and business travelers. Further contributing to tourism's invisibility was its seasonal nature, low-wage jobs, and organizational location outside of government; that is, in tourism and convention bureaus. Tourism had no elevated profile in either economic development or political terms.

If a city wanted to encourage economic growth and increase consumer income through high-wage jobs, the prevailing arguments suggested, it had to focus on manufacturing. If it wanted to increase property tax revenues, retain major retailers, attract office developers, and hold onto its middle class, it had to redevelop its downtown and remove its slums. If it wanted to lure investors and new housholds, it had to have good infrastructure. Tourism was small-scale, seemed incapable of spurring growth, and was isolated within the local economy. Moreover, it had no political clout. Consequently, tourism was neither defined as an 'industry' nor targeted for assistance.

Tourism's ascendence

The subsequent urban restructuring of the 1970s and 1980s put tourism on the economic development agenda. To understand this transformation in status, we first need to understand the forces that expanded and further commodified tourist activities (Urry 1995) after the Second World War. These forces changed the profile of demand for leisure activities and business travel and turned the tourism sector into a more organized, concentrated, and professionalized endeavor. The designation of tourism as an urban industry and its rise to economic development priority were the consequences of a major expansion in the demand for and supply of tourist activities.

Let us begin with the changes in the demand for tourist opportunities. In the 1920s through to the early 1950s, setting aside the disruptions of the depression and the Second World War, leisure activity for most people simply meant not being at work. Days of rest were spent with family or on short trips to the beach or countryside, and long vacations or extensive travel were mainly the prerogative of the rich and the upper-middle class. Until automobiles became widely affordable in the 1920s, being 'a tourist' for most people was a rarity. Even then, the segment of the population with such travel potential was small in comparison to today (Jakle 1985).

A number of factors – rising disposable incomes, the restructuring of work, suburbanization, and rising levels of education – changed all of this. Key to the expansion of tourism demand was the rise in disposable incomes in the 1950s and 1960s. Between 1940 and 1970, disposable personal income more than doubled after controlling for inflation, and the median monthly income of families more than tripled in constant dollars (US Department of Commerce 1975: 225, 296). The economy was prosperous, wages rose, and household incomes increased, initially without many more women having to join their husbands in the work-force. The 'family wage' was the norm in large, mainly unionized corporations and this wage growth was experienced not only by a growing class of professional and white-collar workers, who made up an expanding middle class, but by the high-skilled working class as well.

Additionally, leisure time was becoming more available. The number of days worked per week, and the hours devoted to work, were declining for a good part of the population. In 1909 the average manufacturing worker spent 51 hours per week on the job, and in 1920 47.4 hours per week. By 1950 the average work-week had been reduced to 40.5 hours (US Department of Commerce 1975: 169–70). One study reported a five-hour increase between 1965 and 1985 in free time per week for the average American (Putnam 1996: 38), although other research indicates declining leisure time since 1970 (Schor 1993). In addition, union contracts and the norms of white-collar employment were expanding non-wage benefits, particularly paid vacations of at least two weeks per year. The 'summer vacation' became a widespread practice. Even for white collar, salaried workers whose hours at work exceeded the norms, the incomes at their command meant that when they had 'time off' they could truly 'get away'.

One consequence of higher disposable incomes and more non-work time was increased spending on recreation. Personal expenditures on recreation went from $8.6 billion in 1940 to $28.9 billion in 1970 in constant (1958) dollars, a gain of 337 per cent (US Department of Commerce 1975: 401). Correspondingly, the population was 154 per cent greater, thus indicating rising per capita spending on recreation.

Leisure and recreation, moreover, became conflated. In the last half of the twentieth century, most people approach leisure not as time for contemplation but as an opportunity to be active (Sorkin 1992; Urry 1990; Zukin 1995). By contrast with an earlier period, leisure is now more frenetic. Americans work at leisure, and relaxation outside the house is less a matter of quiet reflection in a natural or familiar setting than of active engagement with nature (e.g. whitewater rafting, hiking) or with others (e.g. playing softball, golf). Travel and sightseeing are central to this modern experience (MacCannell 1976).

Simultaneously, leisure activities consume more goods, services, and places than they did prior to the Second World War. This has paralleled and was part of the commodification of 'community' and the social mobility that characterized post-war suburbanization (Clark, Jr. 1989; Jackson 1985). Suburbanization had (and still has) a strong leisure component linked to backyard barbecues, patios, driveway basketball courts, and other activities that came with increased disposable income and expanded recreational pursuits, all associated with a middle-class lifestyle.

The spread of education also fueled the redefinition of leisure and the demand for tourist activities. From roughly the late nineteenth century onward, travel, particularly when undertaken by a leisure class, has had an educational component (Jakle 1985; MacCannell 1976). To travel was to gain knowledge of other cultures and places. With increasing numbers of people earning college degrees and being inculcated into a middle-class commitment to personal development, leisure became another opportunity for 'becoming educated'. Parents wanted not just to 'get away' but also to 'broaden the horizons' of their children and themselves. Historical sites, foreign places, museums, and exotic environments lured them. College students spent a summer traveling around Europe to 'round out' their education, and although many vacations are sybaritic, others are dedicated to self-enhancement through novel or challenging experiences. Leisure has been re-defined, now commodified and tourist-oriented rather than an opportunity for rest or contemplation at home.

The second major component of the demand for tourist facilities is business travel and the meetings, conferences, conventions and trade shows associated with it (Zelinsky 1994). Here, too, drastic changes occurred just after the Second World War. With the expansion of business services and white-collar employment, the surge of college graduates into the job market, the shift towards information as the *lingua franca* of business, and the growth of large corporations, more and more face-to-face interactions between far-flung corporate colleagues, buyers and sellers, and industry representatives were needed to keep the economy functioning. Trade shows became more ubiquitous, corporate meetings increased in number, and professional organizations added their conferences to the mix. Overall, the demand for meeting and convention facilities went up along with the travel that moves workers from one place of business to another. Between 1964–5 and 1990–1, the number of metropolises with major convention sites went from 167 to

205 (23 per cent), the number of conventions increased by 78 per cent, and the number of participants grew by 121 per cent (Zelinsky 1994: 72).

On the other side of the equation, in the tourist sector itself, changes were underway that resulted in its expansion, diversification, and concentration. One was the professionalization and consolidation of tourist activities (Judd 1995). The post-war period witnessed numerous occupations striving for professional status by establishing organizations, building educational pathways, and redefining their work as a 'career'. All of these activities are geared to developing greater prestige and influence.

Travel agents, convention and visitor bureau personnel, hotel and tour operators, and restauranteurs strove for professional status. The National Tour Association was founded in 1951, the US Travel Data Center was organized in 1973, and in 1976 the National Council of Area and Regional Travel Agencies, the National Council of Travel Attractions, and the National Council of Urban Tourism were established. Originally founded in 1914, the International Association of Convention and Visitors Bureaus was renamed in 1974. The Travel Industry Association of America and the American Society of Travel Agencies were two of a small number of such organizations of a pre Second World War vintage. The former was incorporated in 1941 and the latter in 1931 (McIntosh and Goeldner 1990: 49–51, 97–102; Ritchie and Goeldner 1987).

These organizations turned their attention to expanding markets and capturing new business. As an indicator of the heightened activity in tourism, the number of travel agencies went from 6,700 in 1970 to 29,584 in 1987, a 442 per cent increase (McIntosh and Goeldner 1990: 97–98).

Paralleling this professionalization, and fueled by the availability of capital and new methods of financing large-scale investments, has been a long-term trend towards corporate concentration. Just after the Second World War, Fordism became the major model of the tourism sector (Ioannides 1996b; McIntosh and Goeldner 1990: 103–9). Buses, airlines, cruise lines and railroads have been consolidated. Airline concentration continues, with the top five carriers increasing their market share from 54 per cent in 1985 to 78 per cent in 1991 (Ioannides 1995: 53). Unknown 50 years ago, we now have companies ranging from Walt Disney to Promus that own and manage theme parks and casinos (Harrah's) respectively across the country and internationally.

Small and independently-owned hotels and motels have been absorbed into, or displaced from, the market by hotel chains or large corporations subcontracting with franchisees. In the restaurant field, fast-food (led by McDonald's and Kentucky Fried Chicken) and leisure chain restaurants have proliferated. The bulk of movie theaters are owned by a few major corporations. Combined, Hilton Hotels and Marriott Hotels accounted for 43 per cent of the hotel-chain market share in the United States in 1994 (Lazich 1996: 408). McDonald's in 1995 had 42 per cent of the fast-food hamburger market and Kentucky Fried Chicken had 67 per cent of its market in 1994 (Lazich 1996: 347, 349). The top five movie theater chains (United Artists being the largest) owned 32 per cent of the 26,586 screens in 1994 (Lazich 1996: 456).

Corporations created large tourist destinations: Great Adventure in New Jersey, Sea World in Florida, casino-focused Atlantic City, the Rouse Company's many festival market-places, and resort centers around the country. The national scope and large size of the corporations that manage these destinations enabled and required advertising that tapped regional, national and even international markets. The thrust of the advertising is to think 'tourist destination' when thinking 'vacation'. Equally important, this consolidation of small-scale operations makes the tourist sector itself easier to manage, a factor important for involvement in urban economic development policy.

To the extent that the success of these endeavors depends on regional, national and even global markets, the pivotal issue becomes transportation. To maximize patronage, large tourist destinations depend on travel alternatives that are convenient and relatively inexpensive. In the post-war period, the most significant change has been the spread of and falling cost of air travel and automobile usage. Domestic airline route mileage in operation increased fourfold from 1940 to 1970 (US Department of Commerce 1975: 769). Air travel has been key for visits to national attractions and international destinations – thirty-three times greater in 1970 than in 1940 (US Department of Commerce 1975: 403–4). Additionally, the construction of nearly 50,000 miles of inter-state highways beginning in the 1950s has facilitated regional tourist and business travel (Jakle 1985: 185–98).

Also significant for the expansion of tourism have been the historic preservation and environmental movements. By opposing demolition efforts in the 1950s and 1960s, and passing laws to protect historic sites, preservationists kept intact the historic districts that now attract vacationers to Philadelphia, Boston and Santa Fe (Chang et al. 1996). Preservationists learned that if monies were to be available for preserving sites, visitors had to know about and be attracted to them. Their efforts mimicked those by environmentalists to preserve wilderness areas and create state and national parks that themselves have become tourist destinations. Between 1940 and 1970, the number of national parks, monuments, and historical and military areas went from 157 to 277 (Jakle 1985: 67–79; US Department of Commerce 1975: 396).

Suburbanization also plays a role on the supply-side. The attraction of a city's 'history' is intertwined with the nostalgia (Herron 1993) that suburbanites hold for the places their parents or grandparents left behind. The suburbs are the sites from which tourists and day-trippers are launched.

The professionalization, consolidation, and marketing of tourist sites have all come together in a tourist industry increasingly characterized by cooperation among proprietors and managers. Building on earlier models of coordination, including civic boosterism, actors from the tourism sector have learned to form partnerships among themselves and with governments to tout and enhance their industry. All of this has given the tourist industry greater visibility.

Lastly, framing all of these factors, and particularly central to urban tourism, has been the re-making of local governments from service providers to active participants in local economic affairs (Eisinger 1988; Fainstein 1991). Driven by changes in the economy, local governments have found themselves committed to and even responsible for economic growth. These changes include the shift from economic dependence on heavy industry to a dependence on service functions that are more spatially mobile, new inter-governmental relations that have made local governments more responsible for local economic development, and ideological shifts (with origins in the New Deal) that have generated a Keynesian expectation that governments will generate jobs and provide income supports. Admittedly this is not wholly new, having roots in the civic boosterism of the 1800s. Now, however, this entrepreneurial function is much more pervasive. Consequently, local governments are more susceptible to new opportunities for generating jobs and new investment.

Tourism's new status

The elevation of tourism to economic development status in the cities was the result of two sets of changes, one involving the political economic restructuring of cities and the other changes in the tourism sector itself. As regards urban restructuring, three factors deserve our attention: (1) the chronic fiscal stress that local governments developed during the 1970s; (2) the subsequent search for new sources of economic growth, and (3) the displacement of a production-oriented city economy with a consumption-oriented one. These factors prepared the way for tourism's ascendence, an ascendence that was accelerated by: (4) the redefinition of the tourism sector as a source of economic growth, (5) tourism's alliance with local economic and political elites; and (6) the ability of advocacy groups to create a positive public image for tourism.

Not to be overlooked is the increased professionalization of economic development (Visser and Wright 1996). From approximately 1970 to the present, the economic development policy arena has grown (both at the local and state levels), become more organized and professionalized, and developed an academic component that trains students, re-thinks theory, conducts research, and manages professional journals. Consequently, economic development experts now systematically comb the local economy for new growth opportunities and are less inclined to adopt uncritically political responses to changing economic fortunes (Beauregard 1993c). This benefits emerging sectors such as tourism.

More central to the explanation is the post-1973 restructuring of urban political economies. The early 1970s witnessed an important transformation of local governments that eventually became a major factor in tourism's new status. The recession of 1973–5 brought on a fiscal crisis that has lingered ever since, inaugurating chronic fiscal stress for city governments across the country (Judd and Swanstrom 1994: 307–34). The yearly and problematic balancing of expenditures with revenues has spurred a frenetic and constant search for new sources of tax revenues. Tourism has qualities that make it quite attractive in this regard.

The loss of manufacturing jobs and households in the 1960s and early 1970s diminished the tax base of many cities while the demands on city governments to maintain services, even with shrinking populations in some instances, and undertake remedial action (such as urban redevelopment) strained local expenditures. During this time, the federal government assisted local governments fiscally and contributed to economic development and urban renewal programs. However, the hyper-inflation and high rates of unemployment during the 1973–5 recession, both at double-digit levels, dealt a serious setback to local government finances. The withdrawal of the federal government from fiscal assistance further weakened local fiscal capacities. The squeeze on taxpayers, resulting in the property tax revolts of the 1980s, made it politically difficult to raise taxes or to create new ones. Older, industrial cities with declining tax bases entered a world of chronic revenue shortfalls.

This fiscal stress was exacerbated by the increased inter-governmental competition for taxpayers and investors (Goodman 1979; Lueck 1995). Central city governments not only had to compete with their suburbs for middle-class households and business investment, but also with other cities from other states. The desperation of cities and states for own-source tax revenues re-ignited a contest for economic development that has roots in the mid-1800s. Competition has made it more difficult for all governments. The premise was – and still is – that households and businesses are tax-sensitive and, thus, fewer taxes and lower tax rates will attract and retain them. In the absence of robust economic growth, a condition that most older cities face, this also means less tax revenues.

Tourism offers a way of at least mitigating the tax dilemma, for it holds out the possibility of expanding tax revenues and shifting tax burdens. Taxing tourists and the services that they use means that local officials gain increased revenues without having to tax local residents. Taxes on hotel rooms, car rentals, attendance at tourist attractions and sports events, riverboat gamblers, and airline tickets, the argument goes, fall mainly on non-residents. For the most part, visitors provide the new tax revenues. Tourism-based taxes thus have an attractive political quality.

In addition, the growth of the tourist sector also compels fiscal attention. Policymakers believe that taxing declining industries could drive them into further difficulties and result in additional job and tax loss, but taxing growing industries is unlikely to have these consequences. The fact that tourism's growth has meant job growth, for tourism is labor-intensive, makes tourism even more enticing from an economic development perspective.

Second, with the downsizing of manufacturing (Bluestone and Harrison 1982), particularly apparent during the recessions of the early 1970s and early 1980s, urban policymakers began to search for economic activities to replace this once-dominant source of good jobs and tax revenues (Judd and Swanstrom 1994: 335–66). The decline in manufacturing employment also meant the flight of working-class households from the city. They joined middle-class families who had fled to the suburbs in the 1960s. Politically and fiscally, city governments needed new sources of employment opportunities.

One possibility, of course, was the expanding service sector, not just business services such as finance, insurance, and real estate but also health and educational services. Tax abatement and exemption programs, business subsidies, and property development assistance were mobilized to cut costs for these industries so that they would stay in the cities, pay taxes there, and provide additional jobs. This turn to 'services' encouraged economic development officials to look at the broad range of services produced in the city and led them to hotels, restaurants, convention centers, tourist sites, and related businesses and facilities. Services were seen as the 'next' growth sector and, making virtue out of necessity, a sector that would replace the 'dirty' smokestacks of manufacturing with 'clean' industries.

At the same time, the discovery of tourism as a service industry with growth potential was facilitated by a general belief among economic development experts that cities once dependent on manufacturing needed to diversify their economies (Beauregard 1993a). Adding tourism to the local mix of industries furthered this objective. The portfolio approach to local industry mix does not require that an industry experiences no seasonal or cyclical fluctuations. Rather, the diverse mix of industries will counter-balance seasonality, sensitivity to national business cycles, and secular shifts in external demand. Tourism looks good from this perspective. Because it supplies both local and non-local demand and caters to different groups (business travelers, domestic tourists, foreign tourists, day-trippers and other visitors), it is less likely to respond 'as a whole' to shifting economic tides.

Third, related to this move away from manufacturing was the general shift of cities from places of production to places of consumption (Harvey 1987; Mullins 1991; Zukin 1995). The prevailing opinion was that cities were undergoing a profound transformation in their functional importance, no longer serving as goods producers but now functioning as service providers. This coincided with cities as places distinguished by their educational and cultural opportunities as well as their recreational possibilities. Instead of 'Hog Butcher for the World' (Chicago) or 'Flour City' (Rochester, New York), city leaders turned instead to achieving the status of 'best place to live' or 'best place to do business'.

The affluence of the 1980s, with its consumption-oriented gentrification of inner city neighborhoods and waterfront developments, held out the potential for cities to be places of leisure. Increasingly touted as entertainment centers, cities have invested in sports stadia, bid on international sporting events such as the Olympics or World Cup, and subsidized theater and cultural districts. Downtown redevelopment for the last twenty years has been anchored by retail projects, with the Rouse Company's festival market-places being the model. The relation of this to viewing tourism as central to the city's economy is obvious (Plates 12.1 and 12.2).

Fourth, new perspectives on the urban economy also conferred greater prominence on the tourist sector. Economic development policymakers expanded their understanding of exports to include services, as well as goods, and this enabled tourism to be taken more seriously. In addition, urban economies came to be understood as driven by clusters of industries and not simply by one major export industry acting alone (Doeringer et al. 1987). This turned economic development attention to local linkages among businesses. Once one looked more closely, it became apparent that local tourist activities drew on a range of other businesses from wholesalers to office-supply houses to musicians. Policymakers also discovered the minimal barriers to entry into the sector (thus allowing for numerous small business opportunities), the low cost of job creation in comparison to manufacturing, and the less seasonal nature of urban tourism when compared to resorts and rural areas.

Making the re-positioning of tourism easier was the fact that cities already had many, if not most, of the elements of a tourist sector in place. One did not have to create or attract a whole new industry, as one might have to do with high-technology or TV and movie production. Rather, all cities had hotels and restaurants, tourist sites, and transportation infrastructure – the basics – and even unexploited tourist resources (e.g. historic districts or waterfronts). The adoption of tourism as a growth sector did not entail high initial costs, and even the costs of supportive infrastructure could be defrayed with state and federal grants and tax subsidies.

Fifth, quite important in the economic development ascendence of tourism was the recognition of the relationship between tourism development, property development, and infrastructure. This enabled tourist interests to build alliances with local economic and political élites. Enhancing the tourism sector now means building new hotels, shopping facilities, waterfront promenades, convention centers, international airport terminals, theaters, and sports stadia. It also now requires new roads and bridges, sewer systems, mass-transit improvements, and directional signs. Viewing tourism in this way has enabled advocates to join with property developers and banks, as well as government officials, in support of public–private initiatives. Consequently, the sector has gained economic partners and political allies. Tourism has become a member of the urban growth coalition (Whitt 1988).

Finally, we should not discount the role of advocates in elevating tourism's local status. Professionalization and the development of strong organizations that can speak for the sector are important contributing factors. In addition, researchers have pushed beyond a narrow view of tourism and have cast it as an industry with numerous geographical consequences, business linkages, fiscal implications, and social dimensions (Ashworth 1992; S. Britton 1991; Featherstone 1991; Frederick 1993; Ioannides 1995; Urry 1990 and 1995). By broadening their perspectives, tourism planners have also given the sector more prominence (Haywood 1992; Inskeep 1988). They have begun to address the environmental, land use, and transportation impacts of tourist activities in ways that link tourism to issues that traditionally have enjoyed privileged status in public policy circles. Tourism is now more prominent and intellectually intriguing as a way to understand cities and their economies, and less *ad hoc* in its planning. The influence of these factors, though, still has to be viewed against the backdrop of cities desperate for new sources of jobs and tax revenues, and intent on expanding their responsibility for the health of the local economy.

The last three decades have been fortuitous for tourism. Once nearly invisible in economic development terms, it is today a high priority item on almost all urban agendas.

Conclusions

Tourism's new economic development status, then, is a function of three sets of forces. First, changes in lifestyles and tastes – and the deepening commodification of leisure – expanded the demand for and the supply of tourist activities. Second, urban restructuring compelled city governments to search for new revenue sources and prepared the way for a sector emphasizing leisure, recreation and consumption. Lastly, the professionalization, advocacy and political alliances forged by tourist interests made tourism publicly prominent and economically attractive and gave it access to governmental decision-making. In the world of urban economic development, tourism now has 'priority'.

Can tourism retain this exalted status? One particular response to this question raises doubts. The tourism sector remains sensitive to seasonal demand and changing incomes, themselves sensitive to business cycles, and its employment structure is dominated by low-wage jobs with short career ladders, all of which reduce its attractiveness as a key element in local economic development policy. In addition, the development of a standardized tourist product and widespread adoption of tourism as an economic development target, both within and outside the United States, means heightened competition for the tourist and business travel expenditures. Even though visitor expenditures are growing, they are not growing fast enough to allow all cities to be robust tourist destinations. Further threatening any individual city's tourist sector is market saturation. The lack of coordination across cities, the ease of entry into the industry (made easier by government subsidies), and the aggressiveness of investors (see, for example, riverboat casinos (Deitrick et al. 1999)) mean that market saturation and 'shake-out' are distinct possibilities. As one researcher (Urry 1990: 40) wrote, 'almost every place in the world could well [be] an object of the tourist gaze'.

A different response to the question recognizes that industry characteristics are not immutable and that the future is created, not predicted. Tourism's many proponents can act to avert the potential economic downside. Economic development policymakers might pursue a diversity of tourist activities in order to counteract seasonality and other volatilities. This could be joined with efforts to identify substitutes and complementarities in order to expand and deepen tourism's linkages to the local economy. Industry owners and managers might look to reorganize so as to create longer career ladders while also boosting the wages paid for non-entry-level jobs. Tourism planners might monitor signs of impending saturation and encourage cities to establish niches rather than assume invincibility.

Through such actions, tourism can make itself even more attractive to political officials and economic development policymakers. However, we should not forget that while tourism has emerged from the shadows cast by other industrial sectors, it can just as easily slip back and become, once again, marginal to urban policy.

13

ENTREPRENEURSHIP, SMALL BUSINESS CULTURE AND TOURISM DEVELOPMENT

Gareth Shaw and Allan M. Williams

Introduction

There has been considerable discussion of the role of tourism as a force for economic change (Sinclair and Stabler 1992). Indeed it could be argued that part of the 'rediscovery' of tourism studies by economic geographers has been through its relatively high profile in contributing to employment. This is certainly the case in many mature economies where the success of the tourism industry, particularly in the face of economic restructuring, has enhanced its importance to state policies at both national and local levels (Agarwal 1997). Similarly, tourism has provided the basis for an attractive, alternative growth form for many developing economies in a wide range of geographical settings (Harrison 1994; D. Hall 1991).

However, as we have discussed elsewhere (Shaw and Williams 1994), in spite of considerable debate particularly on developing countries, there is still little consensus as to tourism's role in the process of economic development. The overall picture is somewhat confused by a number of fundamentally different assessments, and is further clouded by the issue of whether tourism actually constitutes an industry and if it does, what the nature of the product is (S. Smith 1988, 1993 and 1994; Leiper 1993b; and see Part A in this book for additional comments). In addition, some authors have raised questions concerning the nature of economic development in tourism economies (Pearce 1989), whilst other complications arise because of the interrelations among the economic, socio-cultural and environmental implications of tourism (McIntosh and Goeldener 1990). In terms of these latter debates, a large part of the difficulty in understanding tourism's impact on economic change stems from a failure to consider fully the nature and role of entrepreneurship (Shaw and Williams 1990). In contrast to many other economic sectors, geographers researching the tourism industry have, apart from a few limited remarks, greatly neglected the issue of entrepreneurial activity (Mathieson and Wall 1982).

Given such general neglect, this chapter sets out to provide a more detailed perspective on entrepreneurship within the tourism industry. We start by outlining the concept of entrepreneurship within tourism and then turn to the key position of the entrepreneurial process in local tourism economies via the resort/product life cycle model. From these more general discussions, the debate moves on to more specific considerations of entrepreneurial activity in different tourism economies.

The concept of entrepreneurship in tourism studies

Entrepreneurship is a fairly complex issue and not always easy to define, with some authors viewing it as a process or a way of behaving (Cunningham and Lischeron 1991). McMullan and Long (1990), amongst others, have taken such ideas further by suggesting that the concept of entrepreneurship can best be understood by de-constructing it into three main overlapping components. These are: *creativity or innovation, risk taking, and managerial or business capabilities*, all of which Echtner (1995) highlighted in her discussion of entrepreneurial training in tourism. However, as will be indicated later in this chapter, it is not always clear whether in many tourism economies it is really possible to identify or discuss entrepreneurs in terms of all three components.

In addition to these functional and behavioral elements, entrepreneurs are also conditioned by particular structures which characterize local and regional economies. Certainly, in the context of general studies of entrepreneurship, as Herbig, Golden and Dunphy (1994) argue, the societal structure of a locale strongly influences the propensity of that society to produce entrepreneurs. Similarly, within economic geography, Cooke (1983) and Massey (1983, 1984) have stressed the role local cultural systems play in economic development. More specifically within the context of tourism, Jafari (1989) has highlighted the importance of what he terms a tourism business culture within some developing countries. It is in the context of local culture that tourism raises more complex issues, since Jafari (1989) argues that tourism entrepreneurs have a role in acting – either directly or indirectly – as brokers within the host community. Such a role is also strongly conditioned by whether entrepreneurs are drawn from the local community, or are external agents. It is at this level that the nature of tourism entrepreneurship becomes deeply entwined with the issues of the socio-cultural impacts of tourism on host communities. According to Jafari, entrepreneurs operate within a business culture, and at the same time, some are also members of the local society. Therefore, given this dual pattern of behavior between the 'ordinary' and 'non-ordinary' worlds, they can act as important brokers between hosts and guests. Such a potential also raises the issue of local versus non-local entrepreneurs, or more especially the role of large transnational corporations in developing tourism economies, a theme we will explore later in this chapter.

The role of local entrepreneurs is also touched upon in the original discussion of the resort cycle by Butler (1980). In this context, he speculated that in the early stages of resort development, as the number of visitors grow and become more regular, some local residents will 'begin to provide facilities primarily or even exclusively for visitors' (Butler 1980: 7). In the original model this phase of development was termed the involvement stage, within which as Din (1992: 11) argues, the postulation is 'that entrepreneurial development is a spontaneous process which occurs as a result of growth in tourist demand' (Figure 13.1). Unfortunately, Butler does not elaborate further on entrepreneurial activity, although Din suggests that it is reasonable to identify two assumptions underlying the involvement stage. The first is that it is local people and not outsiders who take on early entrepreneurial initiatives. This, he argues, is logical on the grounds of the strategic locational advantages gained by those already residing in the locale of tourist growth, stressing the importance of information flows and proximity. Such activity is feasible given the relatively low entry requirements and limited barriers into many parts of the tourism sub-sector, especially those concerned with accommodation. The second assumption is that a sufficient number of local residents are occupationally mobile and, hence, are capable of realizing the

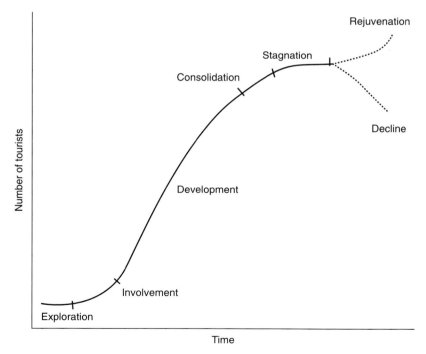

Figure 13.1 The resort cycle model
Source: Butler (1980)

opportunities provided by a growth in tourist numbers. The importance of low entry requirements is again significant since both Butler and Din assume that local people have ready access to capital and have the necessary abilities to function as entrepreneurs. As we shall show later, such conditions can prove to be of critical importance in understanding the actual behavior of many entrepreneurs.

Alternative perspectives on the early stages of resort development suggest that the first major business initiative tends to come from external rather than internal agencies, but according to Gormsen (1981) these rapidly give way to more regionally based entrepreneurs. Certainly evidence from some developing countries, particularly former colonies, is strongly supportive of this type of process, although it is not clear either from Gormsen's or Butler's work, just what constitutes a local entrepreneur. Din (1992) draws attention to Lundgren's (1975) early work which arbitrarily specifies a distance of 25 miles from the resort as constituting 'local', and suggests that this is probably consistent with Butler's original concept of local. But he goes on further to argue that the term local should only refer to those entrepreneurs who reside permanently in or close to the resort. Even this, however, is somewhat problematic when you are investigating resorts where resident entrepreneurs have recently moved to live and work in the area, often bringing with them strong external influences.

As previously mentioned, the balance between external and local entrepreneurs is at the heart of the debate over ownership and control of the tourism economy and, in turn, who benefits from such economic growth. It is for this reason that much of the debate concerning entrepreneurship in tourism has been submerged in a broader agenda about the nature of external control or what some have termed the 'dependency' perspective (S. Britton 1981; Din 1992). This debate is largely set within a political economy context, which generally portrays the development of tourism as a variation of neo-colonial experiences. The question of local versus external control is clearly important not just within developing economies but also in tourist resorts within developed countries, and there are many striking similarities in the debates relating to power structures. As both Hollinshead (1993) and Dann (1996a) argue, in a different context, tourism is about power and truth, and the manipulation of both tends to be through the entrepreneurial system. Control over local power structures, both political and socio-cultural, has a direct influence on the availability of entrepreneurial opportunities in all types of environments. However, in spite of at least superficial similarities, a closer examination of existing studies can best be provided by discussing the detailed role of entrepreneurship in the two separate contexts of developing and developed economies.

Entrepreneurial processes in developing economies

The reliance of some developing economies on international tourism as a growth strategy has been criticized because of its over-dependency on external capital and entrepreneurship (de Kadt 1979). More specifically, Rodenburg (1989) has demonstrated in the case of Bali that large transnational organizations were rather ineffective in increasing foreign exchange earnings and job opportunities. This was mainly due to significant economic leakages, through profit remittances and out-of-region sourcing of material and service inputs. There are also significant structural and geographical dimensions of such dependency. Both S. Britton (1981) and Pearce (1989) have argued that major tourist flows and controls emanate from the developed economies whilst, in the destination country, resort enclaves are created. It is through such spatial networks that transnational organizations operate, and unless they are strongly regulated by governments, only limited economic benefits may accrue to the host communities. Increased globalization trends in the tourist industry, especially amongst the travel and hotel sectors, have accentuated the role of transnational corporations (Cleverdon 1992). As Go and Pine (1995) show with reference to the hotel industry, transnational organizations with their headquarters in the US, Europe and Japan, enjoy strong competitive advantages over their rivals in the host country, many of which may be small to medium-sized independents. Such advantages accrue because of favored access to their domestic markets, accumulated knowledge of clients' tourism practices and preferences, and through linkages with airlines and tour operators.

Elements of external control have been incorporated into some models of tourism development which have sometimes been couched in terms of core–periphery relationships. Thus, Hills and Lundgren (1977) suggest that there are strong hierarchical dimensions in the spatial networks of the tourism industry which are clear expressions of an underlying metropolitan hegemony (Lundgren 1972). In this perspective, major flows of tourists and controls emanate from the developed economies whilst, in the destination country, exclusive resort enclaves are created. However, Din (1992) argues that in stressing the dominance of such metropolitan control via transnational corporations, such a perspective ignores the existence of opportunities for local entrepreneurial activity. Din considers that it is overly pessimistic to view the local community as nothing more than a passive victim in the face of the internationalization of tourism and the globalization of capital. Alternatively, he suggests that some locals have prospects to become entrepreneurs and, in support, cites the case of five hotel groups in Malaysia, all of which grew from local, family businesses to become competitive at an international level. In this sense, the local is more than the outcome of the global, and instead contributes to shaping the global.

We should stress, however, that much of the literature fails to support Din's optimistic view of indigenous growth prospects in most developing countries. In Kenya, for example, foreign ownership accounts for almost 60 per cent of hotel beds, with large transnational hotel groups such as Intercontinental and Hilton International having major developments (Rosemary 1987). Similarly, in the Gambia, the tourist industry is dominated by seventeen large hotels together with a holiday villa complex catering for international visitors (Thompson *et al.* 1995). The impacts of transnational corporations in such small, relatively underdeveloped economies are considerable and include a loss of control over the local tourist industry. In addition, Lea (1988) estimates that the leakage of foreign earnings from these countries can be as high as 50 to 78 per cent. It is not surprising that, under such circumstances, many developing countries remain at best mistrustful of transnational corporations. Such attitudes are clearly reflected in a World Tourism Organization (1985) survey of twenty-two developing economies, which found that most still considered that the disadvantages of such organizations outweigh their benefits.

There are, however, signs of a changing response to this situation by both the transnational corporations themselves and the host countries. Significantly, such responses are manifested at the level of the entrepreneur. Transnational hotels can no longer ignore host community demands for more active involvement in establishing a local tourism agenda (Ritchie 1991). Moreover, as Go and Pine (1995: 346) argue, 'astute hotel firms should become more sensitive to the local community'. They base their views on the wider strategic implications of Barlett and Ghosal's (1989) 'transnational model', which is significant since it allows large hotel corporations to 'balance global reach and local issues' (Go and Pine 1995: 346). Recognition of local concerns has tended to focus on two main resources: environmental and human. We need not dwell on the former here, other than to note that through the International Hotels Environmental Initiative (reported in Williams and Shaw 1996) many hotel corporations have acted upon local environmental issues associated with tourism and started to implement integrated programmes of sustainability. In some hotel corporations, efforts are also being made to improve the management skills of the local work-force. Of particular note are such schemes as that initiated by ITT Sheraton Corporation, which is aimed at developing managerial skills through the launching of an MBA course for senior executives in the Asia-Pacific region of its operations (Go and Pine 1995). It may well be that such high-level training may facilitate local people becoming entrepreneurs by the cross-transfer of skills from the transnational corporation to the local-based tourism industry. It should be stressed, however, that although the potential exists for such a process, it is too early as yet to determine whether it will be successful in this way.

The need to train local people to operate as entrepreneurs is increasingly important in many developing countries. Such local entrepreneurs not only raise the tolerance limits of the host community towards tourist activities (D'Amore 1983), but also as Loucks (1988) suggests, they enhance community stability through empowering local people and giving them a stake in tourism development.

Unfortunately, very few of the international aid programs that have focused on the training of local entrepreneurs in developing countries have initiated schemes within the tourism sector. This is especially surprising on two major counts. First, tourism is a significant growth sector in many of these economies (figures from the World Tourism Organization (1994b) show developing countries had a 25 per cent share of the almost $US300 billion receipts from international tourism in 1992); and second, tourism has far-reaching impacts on other economic sectors through strong backward linkages. As Mathieson and Wall (1982: 82) state, 'there is little doubt that the tourist industry exhibits backward linkages and that external economies have emerged'. More recently, Telfer and Wall (1995) have stressed the need to investigate and strengthen the economic linkages between tourist regions and their hinterlands.

One of the few detailed studies to explore the relationship between inter-firm linkages and tourism entrepreneurship has been the work of Lundgren (1973). He examined the characteristics of tourism-based entrepreneurship associated with the different forms of hotel development in the Caribbean, and proposed a three-stage model of entrepreneurial development. This was based on the supply and demand linkages for food by hotels, and illustrated in its basic form by the example of an island economy. As Figure 13.2 illustrates, stage one is characterized by a new hotel which has no economic linkages with its hinterland, with supplies being imported. Evidence suggests that much of the early development in the Caribbean took the form of large complexes that developed strongly integrated systems with foreign suppliers (Archer 1995; Belisle 1984). This was either because the local agricultural system could not meet the rapid increase in demand or because the hotel was foreign-owned, and had a corporate policy of not using local produce.

Evidence from the developing economies of the Caribbean during the period of rapid tourism development, supports the early dominance of such foreign supplies. For example, Gooding's (1971) study of Barbados estimated that two-thirds of all food eaten by tourists was imported whilst Cazes (1972) recorded similar figures for hotels in Jamaica. Thus, during the rapid period of growth in the 1970s, authors such as H. Brown (1974) thought that such high levels of food imports were a result of domestic agriculture's failure to meet new demand. Latimer (1985), however, has argued against what he considered to be a one-sided perspective, preferring instead to highlight the patchy nature of entrepreneurial activity. He quoted the case of a hotel manager in the Seychelles being unable to obtain locally grown mangoes

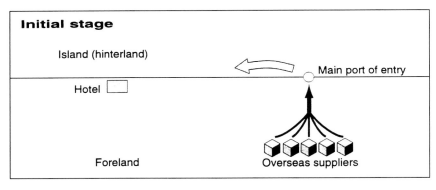

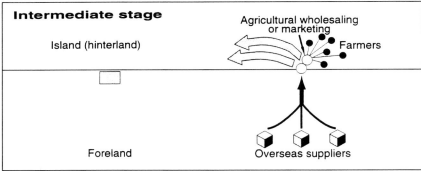

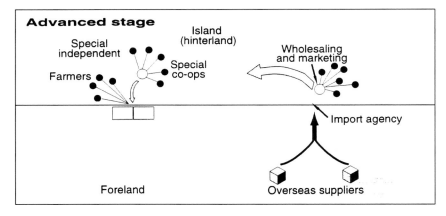

Figure 13.2 Lundgren's model of entrepreneurial development and hotel linkages
Source: Lundgren (1973)

because no effective marketing system existed. There is also evidence to support the view that foreign-owned hotels developed a supply policy around the perceived demands of their guests and showed little or no interest in local suppliers. This is potentially significant in that food represents approximately one-third of the expenditure by tourists in Caribbean economies, and that the degree to which the tourist industry relies on imported food can greatly affect the socio-economic impact of tourism (Belisle 1983). Estimates for the Bahamas suggest that for every American tourist dollar spent, at least eighty cents leaks out of the local economy to purchase imported foodstuffs (Eber 1992). In the same study Belisle also examined food production in the Caribbean and concluded that six main factors influenced hotels' links with local entrepreneurs, namely: most tourists are conservative in their tastes; imported food may be cheaper than that produced locally; hotels will pay more to ensure reliable and high-quality supplies; local food may be processed in unhygienic conditions; hotels may be unaware of locally produced food; and local producers may not know how to contact the hotels. Telfer and Wall (1995: 640) have generalized the situation as one mainly concerned with, 'the interrelationships between entrepreneurs with differing access to resources, and the sizes of enterprises'.

Returning to Lundgren's model, he suggests that local entrepreneurs may develop links with the hotel sector and thereby open up a marketing channel between farmers and hotels (the intermediate stage in Figure 13.2). The final or advanced stage of entrepreneurial activity sees the further expansion of local wholesaling activities in both organizational and technological terms, with the latter involving more cold storage capacity.

Although dated and simplistic, Lundgren's model serves to highlight the backward linkages in the tourist industries' demand for food, and similar economic developments can be explored for other external linkages. Unfortunately, the progression of developing economies through to the advanced stage, as suggested by Lundgren, has rarely been identified. From the limited evidence available, many developing countries appear to be in either the initial or the intermediate stages. For example, Momsen's (1986) work on the small Caribbean islands of St Lucia and Montserrat shows an improvement in the linkages between the hotel sector and local agriculture during the period 1971 to 1983. Thus, in St Lucia in 1971 70 per cent of the value of food consumed by tourists was imported, compared to only 58 per cent by 1983, which highlights the development of improved linkages as a result of better channels of local food distribution.

That entrepreneurship is important in the linkages that tourism forms in developing countries is clear, and it may well be, as Lundgren argues, that such relationships develop in a series of distinctive stages. However, as Mathieson and Wall (1982: 82) point out, the nature of entrepreneurship is contingent: 'Although it is attractive to think of a sequence of developmental stages, the exact pattern of entrepreneurial activity is likely to vary from place to place . . . '. Unfortunately, such geographical variations in the development of entrepreneurship, which can be conceptualized as local contingencies in their formation and operation, have been little researched. One exception has been the general recognition that economic linkages between tourism and local industries depend on: the types of supplier and producers in the local economy; the historical development of tourism within the area (although this is only vaguely conceptualized); and the type of tourist development under construction. A number of these factors are touched on by Momsen's (1986) research which explores variations in tourist patterns and tourism developments in St Lucia and Montserrat. On the former island of St Lucia, hotel investment was predominantly British-based, with 68 per cent of hotels in 1971 being British-owned, and most tourists came from Europe. Montserrat, in contrast, had few European or package tourists, with most visitors coming from the USA. Such tourists also tended to stay twice as long as the average visitor to St Lucia; just over two weeks compared with 6.8 nights. Furthermore, the pattern of development has produced very different structures of tourist accommodation, with villas being more important in Montserrat. Such differential patterns, especially in respect of types of tourist and the structure of hotel development, have strongly influenced the levels of entrepreneurial development within local agricultural systems.

Lundgren (1973) has suggested that the speed of hotel growth is also significant in conditioning local entrepreneurial activity, as gradual development allows time for a succession of infrastructural improvements and creates a gradual increase in visitor demand for local food. However, the experience of much of the Caribbean has not followed this pathway, since most growth was initially rapid over a short period. Indeed, many of the developing countries that experienced early tourism growth fall into this second type of development of the so-called metropolitan hotel model. Under this scenario, growth is rapid, creating an instant demand for large amounts of food products by visitors. Local entrepreneurs are unable to meet such demands and most food is imported, giving rise to the pattern outlined in stage one of Lundgren's model (Figure 13.2).

The recognition and discussion of entrepreneurial activity in tourism development has been partial, being mainly focused on linkages between hotel development and local agriculture. Such a perspective has limited the overall discussion and led to many critical relationships being ignored. Thus, it is important to recognize that there are a range of backward linkages within tourism, and that their utilization very much depends on entrepreneurial processes. As Archer (1995: 924) notes in his work on Bermuda, 'most of the principal tourism-orientated sectors have relatively strong backward linkages, indicating that they are well-integrated into the Bermudan economy'. In this instance, the one major exception was retail-tourism shops which he found had a high propensity to import their goods.

In a broader study of backward linkages, Bond and Ladman (1982) have shown that for Mexico strong linkages could exist between tourism and the local construction industry. Similarly, the demand for local handicrafts may create linkages back to small factories and cottage industries. In addition, external economies can also result from general infrastructural improvements in local transport networks. Conversely, the creation of tourist enclaves significantly hinders the establishment of local linkages and limits the wider economic benefits of tourism, although limited evidence suggests that such enclaves break down over time. However, at this juncture something of a dichotomy arises in the literature on tourism development. Thus, many commentators believe that tourism plays its major role during the early stages of a country's economic development, because in the later stages economies become more diversified following industrialization (Bond and Ladman 1982; Mathieson and Wall 1982; Lea 1988). However, as we have shown, it is precisely in the early stages that the tourism industry very often tends to be oriented away from the local economy. Usually, this is because of poorly developed entrepreneurial skills amongst local businesses involved in food production and distribution. In this context, we can identify a different component or dimension to the so-called 'involvement' stage of the resort cycle. In the established literature, emphasis is placed on local people entering tourism, usually via the accommodation sector. However, much of the evidence from the developing countries suggests that during the early and rapid period of tourism growth (usually in the 1970s), accommodation was provided by external capital. Moreover, the involvement by local businesses was limited as entrepreneurial activity, especially in food production and distribution, often failed to meet demand.

The obvious way of tackling such shortcomings is for the state not only to control tourism growth, but also to initiate and enhance local entrepreneurial processes. As we have seen, there is an increasingly widespread recognition of the need to involve the host community in tourism development, either through the demands for sustainable tourism or in terms of good business practice (Eber 1992; Go and Pine 1995). Such involvement can best be achieved by encouraging and training more indigenous entrepreneurs. However, as Echtner (1995: 128) emphasizes, there are significant 'special issues and problems' which relate to economic, political and social barriers. On closer inspection, it is the social and, to a slightly lesser extent, the economic factors that are particularly distinctive in the case of tourism. By contrast, political instability and stagnation are extremes that tend to characterize a number of the regimes in developing countries and these have broadly the same effects across all economic sectors, and are not tourism specific.

Very few studies of tourism entrepreneurship in developing economies have highlighted the nature of socio-cultural barriers. Thus, Din's (1992) work on the Penang Lungkaui region of Malaysia found that business ownership in multicultural societies is very often controlled by one or possibly two ethnic groups. Such observations are based on the dominance of Chinese enterprises in the accommodation sector of the 'involvement stage' of tourism development. He concluded that, whilst entry requirements remained favorable, it was other factors that dictated the ethnic make-up of tourism entrepreneurs. Of particular note, was the aversion that local (mainly Malays and Indians) had to some tourism activities. In this context host–guest differences, at least for some ethnic groups, were a significant barrier to entrepreneurship. As both Din (1992) and Michaud (1991) argue, ethnic dominance in business tends to be based on historical patterns of ownership, moral philosophy, and/or a strong degree of societal cohesion. In this context, pathways to entrepreneurship in tourism are controlled by a series of non-economic variables, which may combine to marginalize large sectors of a host community.

Such difficulties also relate to economic barriers. Echtner's (1995) review of research in this area concludes that programs designed to improve entrepreneurial skills which have been designed in Western countries tend to be over-complex for developing economies. Given the nature of cultural differences, she argues that entrepreneurial training programs should be place specific so as to optimize the use of local resources. Given the overall lack of financial resources, economic barriers remain an important deterrent to tourism entrepreneurship. Such barriers can be overcome by effective state intervention to provide financial support as well as training programs.

Entrepreneurship and the small business culture – findings from British resorts

At the start of this chapter we drew attention to some of the main charac-teristics of entrepreneurship that have been identified by the literature. As we also explained, most of these studies, even those by economic geographers, have tended to disregard the tourism industry. This is significant, although it does not imply that such literature is unimportant in tourism research but rather that we need to be more selective in its use. For example, much of the literature that is concerned with the retail and service sectors draws attention to the particularities of small business culture, as well as emphasizing strong polarization trends (Kirby 1987), both of which are endemic in many parts of the tourism industry. Moreover, Goffee and Scase's (1983) work on the service industries has contributed substantially to our understanding of different organizational forms of entrepreneurship, whilst at the same time highlighting the tendency of many small businesses to have weak and informal management structures.

There are increasing trends toward polarization in the tourism industry, as shown by the British hotel sector, with major hotels consolidating into large groups and adopting new management styles concerned with brand identity, product quality and consistency. For example, between 1988 and 1992, the number of hotels operated by public companies increased by 3.5 per cent, following a massive 26.1 per cent increase in the period 1986–8 (Harrison and Johnson 1992). At the other end of the spectrum are small to medium-sized operators, whose management abilities and entrepreneurial skills vary tremendously. To an increasing extent, this latter group has come to characterize the serviced-accommodation sector in many British coastal resorts. The reason is that structural trends in the hotel sector have strong spatial characteristics, with the bulk of new hotel investment being in metropolitan areas rather than in traditional coastal resorts. This pattern is partly reflected in the regional growth trends of hotels operated by public companies, as shown in Table 13.1. As can be seen, the main beneficiaries outside London are regions which are not particularly associated with coastal tourism. In contrast, seaside resorts tend to be dominated by independent, owner-occupiers who usually operate small-scale businesses. This is certainly the case in much of the serviced-accommodation sector. For example, in many of the resorts along the Norfolk coast in East Anglia, surveys have found that as much as 84 per cent of accommodation had less than ten bedrooms (Association of District Councils 1993). Similar trends are occurring in the self-catering accommodation sector, which since the mid-1960s has become increasingly important in many coastal resorts. However, from the late 1980s there has been a growing presence of larger companies, using international capital to invest in new holiday complexes away from traditional seaside resorts (Shaw and Williams 1997).

The net result of these trends has been to make many coastal resorts dependent on small businesses, whose operators have questionable entrepreneurial skills. The cycle of decline into which many British resorts appear to be locked into, has been identified by numerous official studies (for a review see Cooper 1997). Surprisingly, however, report after report has failed to identify the critical role played by small-scale business owners in these resort economies. This theme is, in part, highlighted by Urry (1997) who, in addition to drawing attention to the fact that such resorts are no longer

Table 13.1 Main additions by region[a] of English hotels operated by public companies, 1988–92

Region	Additional rooms
South East	1,728
Heart of England	1,466
Thames/Chilterns	1,184
East Midlands	1,123

Source: Modified from Harrison and Johnson (1992)
Note:
a Tourist board regions

fashionable, also argues that a lack of investment has led to poor quality environments.

Our detailed knowledge of entrepreneurial activity within resort areas is unfortunately limited to a handful of studies undertaken during the 1980s. Stallinbrass (1980), researching the Yorkshire resort of Scarborough, provided an early insight into the dynamics of the hotel industry, revealing that around two-thirds of all entrepreneurs lacked any previous experience in this type of business enterprise. Confirmation of such characteristics have been provided by Brown (1987) in his work on Bournemouth and by Shaw and Williams (1990) in a large-scale study of resorts in Cornwall, in south-west England. There are differences in the approaches of these various studies, with Stallinbrass and Brown focusing on the hotel sector, whilst the work on Cornwall investigated entrepreneurship in a wide range of tourism activities. Significantly, however, there are strong similarities in their main findings as summarized in Table 13.2.

These characteristics of small business cultures in British resorts are indicative not of entrepreneurship, but rather of non-entrepreneurship, in that many of the owners show little of those innovative management skills that are defining qualities of the true entrepreneur. Indeed, in the surveys undertaken, most establishments had only rudimentary policies for future growth and few had performed any business planning to carry out their ideas. For example, Brown and Hankinson (1986) found that in the serviced-accommodation sector the future provision of en-suite facilities was the single priority for investment, and they described the enterprises as being dominated by family-oriented aims rather than strictly business ones. In the very smallest businesses, many of the proprietors also worked outside the guesthouse and a number claimed they had purchased the business 'more in order to secure housing for family needs' (Brown 1987: 64). All the studies point to a general lack of professionally managed businesses and very limited product development.

Table 13.2 Summary of main findings of studies on small-scale entrepreneurs in tourism

Main characteristics of businesses
Little or no formal qualifications
Little access to formal sources of capital, family resources most used
Over-reliance on non-paid family labor
Moved into resort, many non-local business operators
Lack of formal business plans and strategies for future growth
No clear marketing strategies, often no marketing takes place
Number of business owners semi-retired, driven by non-economic motive

Source: Based on Stallinbrass (1980); Brown (1987); Shaw and Williams (1990).

If the product was only changing slowly, so also was the promotion and marketing of individual services. The use of marketing strategies was extremely basic and most accommodation units used nothing more than advertising in the tourist guides produced by the local tourist authorities. More recent research has shown very little change from the position in the 1980s. Thus, a survey of small to medium-sized hotels and guesthouses in North Cornwall found that only 10 per cent had prepared marketing plans, while 7 per cent did not market themselves at all. Once again the vast majority (89 per cent) used local-authority-produced guides (Tourism Research Group 1996). Of greater concern for the dynamics of the Cornish tourist industry is that this survey also reveals that many of these small and medium-sized enterprises could not see any value in acquiring any new skills via training courses (Table 13.3).

As Mutch (1995) has shown in relation to IT and small tourism enterprises, the critical variable in conditioning response to change is that of 'top management' or the entrepreneur. In a broader context, the findings of the Cornish surveys of tourism entrepreneurship have identified a number of key factors that appear to condition levels of entrepreneurial activity (Shaw et al. 1987). Of particular importance are the background characteristics of the entrepreneur in terms of age and previous occupational experience, which also relate to the motivations for acquiring or establishing the tourism business. Some of the main findings from the surveys are reproduced for the serviced-accommodation sector in Table 13.4. The main investment rationales tend to cluster around non-economic reasons, such as the 35.8 per cent of respondents who gave environmental factors or 'to come to Cornwall' as their main reason for establishing the business. Of equal importance is the high level of reliance on informal sources of capital, such as family savings (Table 13.4). The study also highlighted linkages between the age of entrepreneurs and sources of funding, as shown in Table 13.5 which shows a greater percentage of people in the 61+ age group using personal savings than any other age category. In one sense this is to be expected since many of these people had taken early retirement to establish a tourism business, bringing with them accrued savings.

Table 13.3 Perceived usefulness of training by owner-managers[a]

Type of training	Very useful	Fairly %	Not useful
In marketing	11	24	65
In business skills	6	20	74
In computing/IT	14	20	66

Source: Tourism Research Group (1996).
Note:
a Based on 200 small businesses in North Cornwall.

Table 13.4 Characteristics of businesses in Cornwall's serviced-accommodation sector

Hotels	%
Ownership	
Individual	85.6
Group	6.0
Limited company	2.4
PLC	1.2
Partner	4.8
Age of owner	
20–30	8.6
31–40	27.2
41–50	32.1
51–60	17.3
61+	13.6
Non-response	1.2
Birthplace (first five regions)	
Cornwall	16.9
South West	8.4
South East	34.9
Midlands	7.2
North West	7.2
Main previous occupations (only top four listed)	
Professional farming	27.3
Secretarial/clerical	16.9
Retailing	13.0
Tourist industry	10.4
Principal sources of capital	
Personal savings	37.1
Family savings	15.7
Bank loan	21.4
Personal savings and bank loan	11.3

Source: Modified from Shaw and Williams (1987) and Shaw *et al.* (1987)

Generalizing from these survey data, it is possible to identify the main controlling factors conditioning levels of entrepreneurial activity in British resorts. Thus, there are two important influences: the age and previous experience of the entrepreneurs, which would be common to most explanations of business activity in a range of economic sectors (Figure 13.3a). However,

Table 13.5 Relationship between age of entrepreneurs and sources of capital in Cornwall

Sources of capital	20–30	31–40	41–50	51–60	61+
			%		
Personal savings	33.3	19.8	41.5	42.0	53.3
Family savings	13.3	12.8	14.1	12.0	16.7
Bank loan	23.3	22.6	9.4	26.0	20.0
Building society and other commercial loans	10.0	12.2	1.9	2.0	0.0
Public agency	0.0	1.1	0.0	0.0	3.3
Personal savings and bank loan	10.0	13.9	18.9	8.0	6.6
Family savings and bank loan	3.3	6.6	5.7	2.0	0.0
Personal savings and building society	0.0	3.3	4.7	6.0	0.0
Pick-up funds	3.3	3.3	0.0	0.0	0.0
Non-response	3.3	4.3	3.8	2.0	0.0
Total (297)	100.0	100.0	100.0	100.0	100.0

Source: Shaw *et al.* (1987)

in tourism the pattern is complicated by the 'motivations' for establishing or acquiring the businesses. Very often these are related to the past experience of the entrepreneurs as tourists, which have helped to shape their views of Cornish resorts and, more importantly, of the tourism industry (Williams *et al.* 1989). The evidence also suggests that there are at least two main types of entrepreneurs. One group is strongly characterized by owners who have moved into resorts, very often for non-economic reasons. Many of these people come with the aid of personal savings, some in semi-retirement, to operate a small hotel or guesthouse. When such motives combine with a lack of experience and an ageing owner, there tends to be a very limited level of entrepreneurial activity. Such owners, we would argue, operate as 'non-entrepreneurs' in that they possess few of the qualities that characterize entrepreneurship (Figure 13.3b). In contrast, a second group is also identifiable, characterized by younger, more economically motivated people who come from more professional (though mainly non-business) backgrounds. These show greater levels of entrepreneurial activity but tend still to be constrained by lack of business skills and capital. These persons can be termed 'constrained entrepreneurs' (Figure 13.3c).

The constrained entrepreneurs obviously represent a resource with potential for development given appropriate access to training and financial help. However, as we have seen, recent surveys have shown a reluctance to

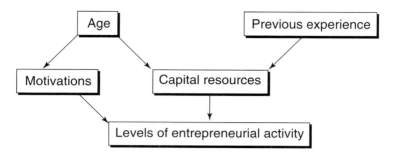

(a) The dynamics of entrepreneurship

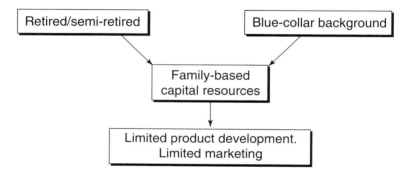

(b) Model of non-entrepreneurship

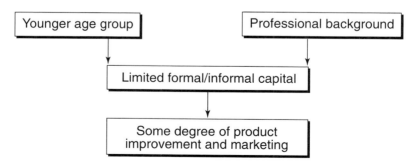

(c) Model of constrained entrepreneurship

Figure 13.3 Dynamics and models of entrepreneurship

improve skills (Table 13.3) by a large number of entrepreneurs. It seems possible that most of those who perceived no real usefulness in training were drawn from the group of so-called 'non-entrepreneurs'. Of the smaller number interested in improving their skills, most are 'constrained entrepreneurs' who would welcome some form of in-house training (Tourism Research Group 1996). However, it seems unlikely from our findings that such entrepreneurs are in a position to utilize any of the DIY strategies suggested by Brownlie (1994).

The limitations of these small-scale entrepreneurs that dominate many British coastal resorts were fully exposed in the late 1980s. At that time, a combination of three main factors (fluctuations in property prices, higher interest rates, and low levels of occupancy) exacerbated the difficulties of small businesses in the accommodation sector (Shaw and Williams 1997). Such conditions produced a climate of high rates of business failures as well as making entrepreneurial activity much more difficult. For example, it has become far more difficult to obtain loans from banks, many of which regard small tourism businesses as highly marginal, and loans can no longer be secured on declining or static property values. Moreover, as we have seen from research in Cornwall, many small tourist businesses are caught in a cycle of under-capitalization, made much worse by the fall in property prices. This may be illustrated by the operating conditions of the coastal resort of Weston-super-Mare in south-west England, which is typical of many medium-sized resorts. Within Weston an analysis of the six smallest hotels for sale during 1991 indicated that they had a value, based on the asking price, of £35,500 per guest room. This compares with a figure of only £19,000 per room for the six largest hotels. If these differences are typical, and there is no evidence to suggest they are not, then it seems clear that smaller businesses will require greater occupancy levels to achieve any reasonable return on capital investment. In reality, however, it is the larger hotels that tend to have higher occupancy levels. Further studies have highlighted the full extent of these problems in other south-west coastal resorts (West Country Tourist Board 1992).

Under these difficult operating conditions it is hardly surprising that many small businesses are under-capitalized and increasingly forced to rely on personal sources of finance, which in turn constrains the entrepreneur. Such conditions have been confirmed in Cornish resorts where research has shown that around 44 per cent of serviced-accommodation businesses had failed to make any material improvements in their businesses since 1980, and that only around 32 per cent had plans for future improvements (Shaw et al. 1987).

Clearly, the limitations of the small business culture – as represented by what we have termed 'non-entrepreneurs' and 'constrained entrepreneurs' – have strongly affected the ability of British coastal resorts to compete. Furthermore, these small-scale entrepreneurs have increased their importance within many small and medium-sized resorts as large-scale capital has moved out in search of more profitable investments in the tourism industry.

Conclusion

It seems clear from our discussion that research, especially by economic geographers, has greatly neglected the role of entrepreneurship in tourism. The literature is divided and limited, with studies being fragmented – in terms of concepts and methods – between different types of economies. Neglect is severest in terms of developed economies and, by contrast, there is increasing recognition of the importance of entrepreneurial processes in developing countries. In these economic environments, it is possible to recognize a growing body of literature that has attempted to consider tourism entrepreneurship within a broader conceptual framework.

It is also possible to recognize commonalities in entrepreneurial processes in most types of tourism economies. Of particular note is the importance of small to medium-sized enterprises within many parts of the tourism industry. These types of enterprise hold a special place in tourism economies for a number of reasons, but most notably because they can provide a strong interface between the host community and tourism. In this respect, entrepreneurial activity not only has an economic dimension but also strong social and cultural elements. Such elements have been readily identified by those studying developing economies in a range of tourism areas. In developed economies, however, the small business culture is being increasingly perceived as a constraining influence on certain areas of tourism growth. Certainly, there is strong evidence to suggest that such small businesses can represent structural weaknesses, leastways in many traditional coastal resorts. Despite these differences, there is some commonality in policies, for training remains a significant strategy for enhancing entrepreneurial processes in local tourism businesses in both developed and developing countries.

In seeking to clarify differing perspectives on tourism entrepreneurship, this chapter has also identified a number of significant areas where further research could be directed. Taken together, these can be seen to represent an agenda for establishment-level research within the economic geography of tourism.

First there is a need to develop stronger lines of conceptual thinking so as to integrate the tourism entrepreneur into the ideas of resort area development. Butler's (1980) original perspective on resort development, as we have shown, hinted at the role of local entrepreneurs in the early or 'involvement' stage of a resort's growth, although much of the discussion was vague and implicitly relied on unresearched ideas. Din (1992) has exposed some of these limitations, but further research is required in this area. At the simplest level, such research may involve no more than the quantification of the 'births', 'deaths' and 'migration' of entrepreneurs and establishments.

However, there is also a need to build on this, and to investigate: the relationships between tourism entrepreneurship and product development and cycles; the conditions under which tourism firms are established and subsequently expand or decline; and the role of social and institutional networks in facilitating entrepreneurship.

In terms of community-level involvement in tourism, local entrepreneurs often hold a key position as studies from developing countries have shown. However, there are important socio-cultural barriers which appear to hinder the involvement of some sectors of the community in the tourist industry. There is scope for further studies of such barriers and to investigate how they may be overcome. Within the context of more developed economies, we again lack research on the role of local entrepreneurs in the local community. Our knowledge of tourism entrepreneurship in these environments is limited to a few studies of coastal resorts, although there is a strong need to extend such research to more diverse geographical environments and tourism sectors. For example, remarkably little is known about entrepreneurship in respect of heritage, cultural or nature-oriented tourism. In this respect, there is also a need to investigate differences between entrepreneurship in the public and private sector, and whether the current vogue for partnerships is stifling or stimulating entrepreneurship. The latter has particular resonance, given the advocacy – in both community and sustainability approaches – that the local community should be one of the partners in developing tourism.

Turning to policy, a number of key research areas can be identified, including the training requirements of small-scale tourism businesses, the most effective financial frameworks for stimulating entrepreneurship (whether public sector grants or venture capital), and the effectiveness of such asso-ciational activities as joint-marketing or purchasing groups. While there is a need to develop such research in most regions and countries, there is also a need to understand how the contingencies of place contribute to shaping the processes of entrepreneurship. As we have seen, there is considerable diversity in entrepreneurship even in the British coastal resorts. This reinforces the need, therefore, for policy-related research to be advanced in tandem with an attempt to deepen our theoretical understanding of tourism entrepreneurship.

14

TOURISM IN THE THIRD ITALY

Labor and social–business networks

Paul Mackun

Introduction

Recently, geographers have advocated closer integration of tourism research into mainstream economic geography research (S. Britton 1991; Shaw and Williams 1994; Ioannides 1995; Ioannides and Debbage 1997). The growth of the Third Italy, and the economic, social and political conditions behind its expansion, has been one of the leading lines of research in this context (Malecki 1995; Piore and Sabel 1984). Until now, however, few geographers have used findings derived from studies on the development of industrial districts, and the flexible means of production that the firms within these districts employ to assess tourism.

This chapter seeks to remedy this situation. It examines previous research on changes in the manufacturing production methods and the manifestation of these changes in the so-called 'Third Italy' (Bagnasco's (1977) differentiation of the regions of north-central Italy from a wealthy north, centered on the cities of Turin, Genoa, and Milan, with the emphasis on heavy industry, to an economically less-developed south, the Mezzogiorno, dependent primarily upon agriculture) (Figure 14.1). This chapter explores the patterns of agricultural ownership (e.g. share-cropping agreements), the political climate in the local governmental units (e.g. the provision of assistance to small businesses), and the social–business relationships (e.g. the development of formal and informal networks of individuals and businesses) that have contributed to the success of several manufacturing industries in the Third Italy. It will then investigate a prominent tourism district within the Third Italy, the Province of Rimini, comparing and contrasting the characteristics of manufacturing labor in the area with characteristics of tourism-related labor in the Third Italy.

Province of Rimini

Regions of the Third Italy
1 Trentino-Alto Adige
2 Veneto
3 Friuli-Venezia Giula
4 Emilia-Romagna
5 Tuscany
6 Umbria
7 Marche

0   200   400   60(

Kilometers

Figure 14.1 Italy

The Third Italy

Over the past few decades, industrial geographers have studied the changing organization of production and labor in manufacturing in industrialized countries; a major focus has been the apparent shift from Fordism to flexible production. Fordism, the dominant system of production in most western economies from the 1920s until the 1970s, is associated with 'very large production units using assembly-line manufacturing techniques and producing a large number of standardized products for mass consumption' (Dicken 1992: 116). In Fordism's heyday, agreements between management and the unions facilitated high productivity for the companies as well as high wages and job security for the workers, the so-called 'labor accord' (Dicken 1992).

In the 1970s, as many industrialized nations underwent a long period of high inflation and high unemployment, Fordism started to show signs of stress (Harvey 1989c). The system of production began to exhibit some of the signs of the 'internal contradictions of capitalism': the 'rigidities of long-term and large-scale fixed capital invested for mass production systems (in turn) predicated upon rigidity in product design and lack of flexibility in relation to mass markets' (Harvey 1989c: 106). Academics and public officials cast a critical eye on the system and began to investigate alternative forms of production in industries such as ceramic engineering, hide manufacturing, and shoe and clothing production (Scott 1992); one region that attracted considerable attention was the Third Italy (Figure 14.1) (Bianchini 1991; Scott 1992; Harrison 1994; Piore and Sabel 1984). Researchers began an examination of this part of Italy, composed of the regions of Umbria, Tuscany, Emilia-Romagna, Marche, Veneto, Fruili-Venezia Guilia, and Trentino-Alto Adige (Bagnasco 1977; Scott 1992), because of the record of success of its small and medium-sized manufacturing firms. They also examined the reasons behind this success and the possibility of replication in other parts of the industrialized world.

The Third Italy experienced economic expansion at a time when other parts of Italy, as well as many other industrialized nations, were undergoing economic decline. This expansion, however, was not fueled by large firms employing Fordist methods of production, but instead by small, specialized establishments that manufactured limited quantities of highly variable goods ranging from ceramics to textiles (Capecchi 1990; Scott 1992). The industrial districts of the Third Italy have become more than simply manufacturing entities; they have become economic, social, and political entities that have engaged in very specialized production. Producers in the region have demonstrated that they are better equipped than their Fordist competitors to cater to the special needs of individual customers (Piore and Sabel 1984). In fact, this flexible system has expressed itself through different manifestations, including numerical and functional flexibility, and computer technology.

Under numerical flexibility, management adjusts the number of workers hired at any given time according to the level of output demanded while, in functional flexibility, management adjusts the assignment of tasks to a set number of employees according to the level of output demanded (Dicken 1992; see also Chapter 6 in this book). In the case of the Third Italy, firms have relied upon both functional and numerical flexibility (DiBella and

DeNicolo 1985). Moreover, firms in the Third Italy and elsewhere have benefitted from the implementation of information technologies, particularly computers, in manufacturing production that facilitate faster and more accurate responses to changes in demand. Indeed, these information technologies have not only improved production processes but also benefitted the distribution of final products (Harrington and Warf 1995).

One prominent Third Italy success story identified by scholars is Benetton, a clothing and textile corporation that began as a small family business and has expanded its franchises into several countries. It benefitted from the advantages of a family that was well-versed in business administration and entrepreneurship, a relatively peaceful labor environment in the region, and financial assistance and technological advice from local governmental units (Harrison 1994). Its individual franchises maintain contact with the corporation's headquarters through an advanced computer system, enabling rapid transfer of information on supply and demand of specific products (Harrison 1994). It is, in turn, representative of successful companies in the Third Italy that are characterized by 'the quality and caliber of managerial personnel . . . competitive strategies, innovations, and the deployment of new technologies' (Poon 1990: 109).

Beyond just describing the growth of manufacturing in the Third Italy, such as Benetton, the researchers have also identified significant causes for the region's distinctive development, including the agricultural and economic antecedents to entrepreneurship, close social networks, and a distinctive political structure; these reasons may shed some light on the provision of tourism services in the Third Italy as well. Unlike many of the landholders of southern Italy who operated large plantations in a semi-feudal manner (Scott 1992), the land-owners of north-central Italy initiated sharecropping agreements. The agreements placed a high level of responsibility on the sharecroppers, both men and women, rewarding individuals with the financial talent and insight to manage their affairs wisely, and motivating the sharecroppers to develop skills such as basic financial accounting and entrepreneurship. Some of the more enterprising farmers, with the requisite skills, helped to launch a crafts industry, relying upon high-quality production as well as, in the case of firms such as Benetton, sophisticated advertising campaigns in both the print and electronic media.

Along with the nurturing of this early entrepreneurship, the development of strong social networks and the nature of social attitudes among the entrepreneurs helped to create a distinctive local labor market that was attractive to investors (Scott 1992). The employers had access to the advantages of extended families, offering greater opportunity for the flexible use of labor, as well as the benefits of a tight-knit community that could provide financial backing to family, friends, and neighbors for business development and expansion. In addition, the attitudes of the employees towards work and towards relations between the sexes differed from other parts of Italy.

The degree of fluidity of individuals between employer and employee status, and a generally more positive view of work by employees, led to an absence of the hostility between labor and management that was found in many communities in the western parts of the country (e.g. Rome) (Capecchi 1990). At the same time, according to Capecchi (1990: 27), women in this area of Italy 'had more power than in other regions of Italy . . . they [were involved in entrepreneurial activity] in [industries such as] clothing and ceramics'.

Political structure and behavior have also played a critical role. Through the twentieth century, many of the communities in the Third Italy, most noticeably in Emilia-Romagna, elected left-of-center governments. Largely shut-out of national politics, officials in these governments concentrated their power over local areas by seeking 'to defend the local society from the penetration of the national state' and 'from the excesses of capitalism' (Bianchini 1991: 338). In order to boost already strong ties to crafts-manufacturers in the region, they actively encouraged the growth of unionization and cooperatives (Bianchini 1991; Hine 1993). Moreover, they used fiscal policy directly to assist the small and medium-sized firms, offering the reduction or elimination of a variety of taxes to selected companies and a relaxation of labor laws for firms with fewer than fifteen employees. Furthermore, they funded a series of technical institutes to create a skilled labor force that would be competitive with other regions of Italy as well as with other industrialized countries (Capecchi 1990; Putnam 1993).

In recent decades, scholars have been able to identify and explain the economic, social, and political factors behind the widespread adoption of flexible forms of production in manufacturing in the Third Italy. Many firms have experienced increased profit margins and an expansion of their labor force, thereby benefitting the communities in which they are located and helping to provide some of the resources to support the growth of other activities such as tourism (*Quaderni* 1994).

The tourism labor market

A handful of academics have noted that the research methods relating to manufacturing can be applied to analyses of the tourism sector (Mullins 1991; Poon 1988a and 1993; Urry 1990). Thus far, however, few scholars have researched labor characteristics in the tourist industry (Ioannides 1995; Shaw and Williams 1994). In particular, there is a paucity of literature relating to the impacts of successful manufacturing practices on the tourism sector.

Nevertheless, the industrial research on the Third Italy has begun to stimulate similar lines of research relating to tourism. After all, 'tourism is one of the largest industrial complexes and items of consumption in modern Western economies' (S. Britton 1991: 451). Each year, world-wide, over 500 million individuals travel overseas, primarily for leisure purposes. The visitors engage in tourism activities that create over 175 million jobs and generate over $6.2 trillion revenue, composing almost 6 per cent of world-wide Gross National Product (GNP) (Lundberg *et al.* 1995).

Thus, in belated recognition of tourism's importance, sociologists, geographers, and other social scientists are beginning to examine characteristics of the provision of services in the tourism sector in a framework similar to that of manufacturing studies of the Third Italy. These characteristics include the greater emphasis on the satisfaction of individual consumer needs, the changing function of labor, the organization of employment (including its social composition), and the role of the state in the tourism sector (Urry 1990; S. Britton 1991; Bagguley 1990; Kinnaird *et al.* 1994; Breathnach *et al.* 1994).

Just as an increased attention to the tastes and preferences of individual consumer demand has gained the ascendancy in selected aspects of modern manufacturing, so too has consumer demand greatly influenced the character of services provided by modern tourism (Poon 1988a). Employment in tourism often necessitates an especially close relationship between the 'producer' and 'consumer' of tourism services, sometimes assuming even more significance than in other service industries. This is due, at least in part, to the unstated aims of modern tourism – the tourist's 'search for the unfamiliar and the unconventional: peoples, places, sights, behaviors, and settings into a commercialized and institutionalized system constructed to satisfy demand for these experiences' (S. Britton 1991: 454).

The customers are seeking a pleasurable experience that is both different and more exhilarating than their daily, mundane experiences at home; one that justifies their expenditure of time and resources as well as their choice of a certain locality and a particular establishment within that locality (Shaw and Williams 1994; C.M. Hall 1994; Poon 1994). The consumers have preconceptions of 'proper' behavior for tourism employees. The first contact is often the most important: 'a moment of truth' (Urry 1990: 71) exists in almost every transaction where the customers either realize satisfaction or dissatisfaction with the service that they are receiving. In order to pass this 'moment of truth', the employees must perform a variety of 'emotional' tasks and do so in a genuine manner: smiling, offering a warm greeting, engaging in small talk, or experiencing empathy. They must also conform to the social expectations of the purchaser of tourism services concerning the physical appearance and behavioral patterns of the employees. The consumer preferences and tastes, therefore, influence the skills that employees value and the age, gender, and racial compositions of individuals chosen to fill these positions (Urry 1990).

While Urry's (1990) research has helped to expand one facet of our understanding of tourism production, and begins to position the study of this industry within the sphere of social and economic geography, tourism still remains severely under-represented in industrial geography research, particularly in labor studies (S. Britton 1991). Many governments, both national and local, invest tremendous amounts of human and financial resources into the development, organization, and promotion of the tourist industry under their jurisdictions, often with the express intent of facilitating greater equity in their communities and regions, and assisting individuals who have been marginalized in the economic processes. These efforts can be improved by a clearer understanding of employment characteristics and patterns in the sector, as well as the potential roles played by economic, social, and political forces on the industry in which they are employed (S. Britton 1991; Ioannides 1995; Shaw and Williams 1994).

Hennessy (1994) and Breathnach et al. (1994) have conducted empirical studies in south-west Britain and Ireland, respectively, that consider some of the economic, political, and social consequences of consumer demand and tourism provision within the context of gender. Hennessy (1994) surveyed the structure and characteristics of tourism in Cornwall, England. Because of the significant differences in the number of tourists between 'peak' and 'non-peak' seasons, many employers in the tourism sector in this area have emphasized one aspect of numerical flexible production, seasonal and part-time employment. Hennessy (1994: 43) found that most positions are part-time and over two-thirds are either classified as seasonal (1–24 weeks) or short-term (25–40 weeks). Of those belonging to this part-time classification, over two-thirds are women; the majority are married and have at least one child. Hennessy documented significant gender differences when positions in tourism were disaggregated by individual job categories. In hotels and accommodations, men were represented in large percentages in both unskilled positions such as portering, as well as skilled positions such as management and personnel administration, while women tended to be concentrated only in low-skilled positions such as waiting tables and cleaning rooms.

Breathnach et al. (1994) used governmental data from both Ireland and the European Union to investigate many of the same issues for the Republic of Ireland, while focusing more closely on the efficacy of the role of the state in the process. Because of its historically peripheral status in the European economy, and its difficulty in finding employment for many of its citizens, the Irish government has recently looked to the development of tourism as a potential catalyst for employment creation; its stated goal is to create as least 25,000 new jobs before the end of the century. According to Breathnach et al. (1994), the Irish government's recent investments in the growth of the tourism sector have only expanded seasonal and low-paying employment, thereby actually entrenching unequal access to positions and offering relatively few positive benefits for the majority of the industry's new employees.

Others, however, take a somewhat different stand on the role of tourism and, indeed, offer cautious optimism for the community as well as the individual. Referring to the contribution of tourism in cities of the United States, for example, Shaw and Williams (1994: 221) argue that 'even if the numbers and quality of [tourism] jobs are debatable, supporters of inner-city tourism point to its role in helping to refurbish the urban environment'. Tourism provides the opportunity for a locality to expand its economic base and enrich employment prospects for its citizens. If tourism employment is accompanied by access to training (e.g. hotel management, logistic operators, translators), it can offer the chance for advancement within the industry.

The previous studies demonstrate the need to better understand labor issues within the context of a particular tourism district. The chapter now turns its attention to a case study of tourism development in the Third Italy, focusing mainly on labor and entrepreneurial issues.

Tourism in the Province of Rimini

The experiences of manufacturing in the Third Italy have directly and indirectly stimulated studies on the decline of Fordism and the expansion of flexible production in several industries, particularly in crafts and high-technology manufacturing (Scott 1992). They have also led to the investigation of the potential for replication of the Third Italy's success in other parts of Europe (e.g. Denmark, southern Germany, and southern England) and in other parts of the world (e.g. selected regions of the United States and of East Asia) (Amin and Robins 1990). Little, if any, attention, however, has been paid to the characteristics of the tourist industry within the Third Italy itself. The region contains a string of sea resorts along the Romagna Riveria, the most important district of which is the Province of Rimini (*Quaderni* 1994; Mackun 1976) (Figure 14.2). The province offers a type of regional laboratory to study characteristics of labor and the state's role in tourism provision. It is located in the south-eastern corner of Emilia-Romagna, a region containing approximately four million residents that has been one of the primary beneficiaries of the economic rise of the Third Italy (Eurostat 1993). During the 1990s this province and other districts within the region have enjoyed a substantial rise in household income, an unemployment rate less than half the Italian average, and a steady increase in the participation of women in the workforce (Eurostat 1993).

The tourism sector has been an integral part of the Rimini economy for over a century. Local legend states that formal tourism began in 1823, with a visit by Luciano Bonaparte, brother of the French emperor. For the next sixty years, the area, containing mostly farmers and fishermen, received a small stream of outside visitors, primarily foreign and wealthy (DiBella and DeNicolo 1985). As in the case of the region's manufacturing industry, the actions of the state combined with individual 'entrepreneurial spirit' gave tourism an important early impetus. In 1860 the Bologna–Ancona railway

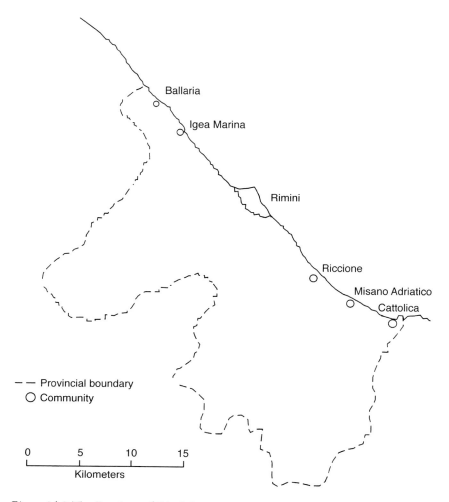

Figure 14.2 The Province of Rimini

was completed, connecting the Rimini area with major metropolitan centres (DiBella and DeNicolo 1985). Positive 'word-of-mouth' advertising increased the flow of visitors attracting, for decades afterwards, Italians from other parts of the country as well as a small stream of wealthy outsiders, primarily from present-day Germany, Austria, Hungary, and the Czech Republic (DiBella and DeNicolo 1985; *Quaderni* 1994). Some farmers and fishermen of the area saw an opportunity to acquire extra income by offering rooms in their residence to visitors (interview with Monetti family in 1995). By the turn of the century, national and local governments once again assisted the growth of tourism through the construction of dozens of hotels in the area, and the redevelopment of parks, streets, and public transit (DiBella and DeNicolo 1985).

The Province of Rimini is exceptionally well endowed with physical and cultural elements that have facilitated the development of the tourist industry. The province possesses miles of well-maintained sandy beaches along waters that are warm for most of the year. Many of the seaside resorts, particularly Rimini and Riccione, emphasize their nightlife offerings in an effort to attract individuals in their twenties and thirties. For families and children, the resorts offer recreational parks and an aquarium featuring an aqua-theater of dolphin performances. The larger communities have built several museums and art galleries that contain artifacts found in local archeological sites dating as far back as the Iron Age.

Further inland, the area contains several hills and valleys suitable for outdoor activities ranging from hiking to agritourism (Rimini & Co. 1991). Within the past decade, the province has begun to serve as the site of business meetings and conferences; the area has invested considerable resources in providing meeting and parking facilities necessary for such events. Furthermore, local businesses have begun to offer copying and duplicating services, and slide and transparency preparation (Rimini & Co. 1991).

The many amenities and services offered in the area attract a diverse group of both domestic and foreign tourists. Each year, almost two million visitors stay at hotels in the province. Between 20 and 30 per cent of these visitors are foreigners. Of the foreign tourists, the largest proportion come from Germany (between 30 and 36 per cent of the foreigners). Those from France and Switzerland generally occupy the second and third ranks (composing approximately 8 per cent each) (Statistica del Turismo 1991 and 1995).

The employers and the establishments of the province's resort areas cater for a wide range of visitors – from single men and women to young families and senior citizens (*Quaderni* 1994). The numbers and diversity of these visitors have created demands for employees in the tourism sector who can cater for a variety of languages and cultures.

The tourism labor market in the Province of Rimini: data and analysis

Although data on the labor markets in the area have been traditionally sparse, recent information from the *XIII Censimento Generale Della Populazione* (the 13th General Census of the Population) and the CGIL (Confederazione Generale Italiana del Lavoro Rimini – General Confederation of Labor of Rimini) offers researchers new insights into a prime tourism market, particularly the accommodations sector. By documenting the characteristics of the establishment in which the employees work, the attributes of the employees themselves, the development of business networks among tourism employers, and the impact of local and national government assistance on the quantity and quality of employment in tourism, the data enable scholars to gain a clearer picture of labor in the sector, and to gauge the degree of similarity between characteristics of employment in tourism and employment in manufacturing districts in the area.

Similar to patterns of ownership of manufacturing establishments in surrounding districts of Emilia Romagna – and in some contrast to the situation in tourism areas studied in other parts of Europe – the ownership of hotels, for example, is controlled to a large degree by local families: 43 per cent of hotels in the district are exclusively family-owned, while another 18 per cent are family-owned in partnership with outside investors (*XIII Censimento Generale Della Populazione* 1991) (Table 14.1). In addition, many of the establishments are relatively small. Over one-half of hotels have fewer than forty rooms (*XIII Censimento Generale Della Populazione* 1991). As one might find in other European countries such as Great Britain and Greece, the 'mom and pop' businesses in the Third Italy form as critical a backbone to tourism in the area as they do for manufacturing (Martinese and Gregoretti 1996).

A 1992 survey by the CGIL of workers in the accommodations sector offers further data on employees who work for these establishments, especially place of origin, educational achievement, and gender composition. The vast majority (94 per cent) of the workers reside in the district, with small percentages from neighboring districts. Unlike tourism employees in other parts of Europe, few workers are immigrants, perhaps reflecting the traditional emphasis on local labor in employment in most industries of the region – small, family-owned establishments tend to attract employees from the immediate area. The workers' education levels are mixed; just slightly under half of the respondents had completed upper secondary schooling, but this percentage has been on the rise. Approximately two-thirds of the participants are women, the majority (55 per cent) of whom are between the ages of 30–49 and have children (*Quaderni* 1994); this latter characteristic is similar to findings from other areas in Europe.

As mentioned earlier, the industrial geography literature demonstrates that most manufacturing firms in the Third Italy employ a small number of

Table 14.1 Ownership of tourist accommodations in the Province of Rimini

Ownership	%
Solely family-owned	43
Family-owned in partnership with outside investor	19
No family ownership	38
Total	100

Source: Based on *Quaderni* (1994)

workers. At one level, the tourism sector appears to show the same pattern. In the early 1990s approximately half (50.4 per cent) of the firms employed five or fewer employees, and over 85 per cent employed ten or fewer employees – comparable to their counterparts in manufacturing in the Third Italy (Martinese and Gregoretti 1996). At the same time, however, larger hotels are important: over 40 per cent of the employees work for establishments that have eleven or more employees (Martinese and Gregoretti 1996).

In contrast to manufacturing, the employees in tourism tend to operate on a seasonal basis; over two-thirds of the positions are less than year-round. Gender differences express themselves in the degree of seasonal employment: 76 per cent of women and 55 per cent of men work in the tourism sector between 2.5–4 months, while 19 per cent of women and 42 per cent of men work in tourism four months or more (*Quaderni* 1994). This is fairly consistent with other tourism areas in Europe (Breathnach *et al.* 1994; *Quaderni* 1994). Male participants in the sector are far more likely to view their employment as the primary source of income in their households than the female participants.

Gender differences become prominent when employment in the accommodations sector of the tourist industry is broken down by job category. Women compose the majority of personnel administrators, chefs/cooks, and waiters/waitresses, while men compose high percentages of bar-tenders, porters, and maintenance workers (*XIII Censimento Generale Della Populazione* 1991) (Table 14.2). This contrasts somewhat from other tourism areas studied in Europe. In north-west Europe researchers found that women dominated the cleaning and food service positions while men were represented in large percentages in both unskilled positions, such as portering, as well as skilled positions, such as management and personnel administration (Breathnach *et al.* 1994). Although the predominance of men in physically oriented occupations in the Province of Rimini is similar to that in other

Table 14.2 The distribution of job categories in the accommodations sector by gender

Job category	Female	Male	Total
		%	
Managers	47	53	100
Personnel administrators	65	35	100
Cooks and chefs	60	40	100
Chambermaids and waiters/waitresses	60	40	100
Bar-tenders	23	67	100
Porters	11	89	100
Maintenance	5	95	100

Source: Based on the *XIII Censimento Generale Della Populazione* (1991) and *Quaderni* (1994).

regions, some differences exist. These include the relative parity between men and women within managerial positions, and the higher proportion of women represented in food preparation tasks (cooks and chefs). This relative equity appears to differ significantly from other tourism areas studied (Breathnach *et al*. 1994).

As for the operators of small and large hotels, many exhibit an ever-growing interest in business networking. In increasing numbers, they are joining community and district organizations of public enterprises and private individuals committed to coordinating efforts to plan for the future of tourism in the province. In these organizations they are cooperating in several areas. They are promoting the Rimini district to potential domestic and foreign visitors using attractive images of the beaches (the coast) and the towns (the interior) in a variety of media to appeal to a host of potential visitors, ranging from singles to young families and retired couples (*Quaderni* 1994). In addition, they are also conducting market research and disseminating the advantages of traveling to their communities to travel agencies both inside and outside Italy (Rimini & Co. 1991). They also communicate their collective needs to governments at the community, provincial, and regional levels. The close-knit nature of the communities (extended families and neighbors) has increased the strength of these groups. The operators of hotels and other tourism services are beginning to value social–business networks in the way that individuals in small-scale manufacturing in the region have done for generations.

In the tourist industry itself, as in the case of manufacturing, the role of the state cannot be ignored. Over the past few years, the district and regional governments – responding to the previously mentioned, increased attention of tourists to the level of skills displayed by tourist providers, as well as to the influence of the business networks – have invested heavily in training and skill-improvement for both prospective and current employees in the tourism sector. The local and regional governments have created three training institutes, two devoted to employment in hotels and one for general training in tourism employment. In the first two institutes, students learn skills related to management and food service (*Quaderni* 1994). The third institute not only trains students for employment in the accommodations sector, but also offers courses in tourism economics, planning, and promotion. Graduates tend to acquire positions outside the accommodations sector. The instruction does not end with prospective employees, however. The local governments support 'continuing-education' programs for current employees of the tourism sector. These courses are not limited to training for jobs in the accommodations sector alone, but also include training in activities as wide-ranging as interpreters, logistic managers, computer technicians, and tourism promotion agents; they are preparing individuals to cater to the specialized needs of an increasingly diverse group of visitors who themselves have higher expectations for their tourism experience (*Quaderni* 1994). As with vocational education in manufacturing-related sectors, the training provides the opportunity for better-paying (and more prestigious) positions – a chance for individuals to improve their lives while being employed in a rapidly growing sector of the economy.

The examination of tourism in the Province of Rimini, based on the evaluation of quantitative data on tourism employees in the district as well as the analysis of the significant factors in the development and growth of the industry, offers one means of positioning tourism, especially the study of labor in tourism, figuratively and literally within labor-related research in economic geography. It also demonstrates the importance of several economic, social, and political factors similar to both manufacturing and tourism in the region while, at the same time, identifying new findings specific to labor in tourism.

Discussion and conclusion

Sociologists, geographers and other social scientists have recently begun to study tourism as an industry. These studies examine the provider–consumer relationship in tourism, the nature of employment for different sectors of society, and the role performed by the public sector. This chapter has reviewed selected issues in economic geography, particularly the role of flexible specialization in restructuring of manufacturing practices in the Third Italy. Throughout, it has been argued that this paradigm can be applicable to studies of tourism within this region.

Previous research on small-firm manufacturing in the Third Italy offers an excellent example of the successful combination of private entrepreneurship, social–business networks, and government intervention in a system that has been judged to be successful in meeting the needs of its customers. In this study, data from a tourism area, the Province of Rimini, situated in the heart of the Third Italy and offering a wide array of amenities to visitors, have enabled a closer examination of employment characteristics in tourism as well as allowing comparisons with manufacturing.

The location of tourism within this successful area of manufacturing illustrates several interconnections, as well as a few differences with the latter. The success of manufacturing in the Third Italy has helped to provide the financial resources needed to support the development of tourism. The entrepreneurial spirit, conspicuous in the development of manufacturing districts, is just as evident in the transformation of several small coastal villages into tourism centers popular with Italians and foreigners alike. In particular, the long-standing social networks have assisted a wide variety of small-scale industries in communities throughout the Third Italy. The expanding organizations of accommodations and other tourism service providers have helped to pool financial resources, share questions and concerns, lobby local and regional government agencies, and coordinate marketing and advertising efforts.

Both direct and indirect government assistance has encouraged tourism. Financial assistance enables businesses to begin and to expand, while the sponsoring of vocational education in tourism provides highly trained employees who can propel the growth of the industry. Of course some differences also exist between tourism and manufacturing in the area, particularly the degree of seasonality and, in some important areas, the gender composition within individual tourism positions. Nevertheless, the elements inherent in the region's success of manufacturing have also played a significant role in the evolution of tourism. The further incorporation of economic geography theory and practice offers even further opportunities for the advancement of tourism provision and other tourism-related issues.

Acknowledgments

The author wishes to acknowledge gratefully the cooperation of tourism officials in Rimini, Riccione, and Cattolica (Provincia di Rimini, Italy), and offer special thanks for the assistance of the Monetti family, owners of the Hotel Europa-Monetti, Cattolica, and long-time residents of the Provincia di Rimini. This material is based upon work supported under a National Science Foundation Graduate Research Fellowship.

Part E

CYCLES AND INNOVATIONS

15

ECONOMIC BUSINESS CYCLES AND THE TOURISM LIFE-CYCLE CONCEPT

Michael Haywood

Boom–bust cycles and tourism

Economies move in cycles from boom to slump, and occasionally to bust. Demand for travel and tourism services coincides with these economic cycles. Importantly, the entire travel and tourism industry displays extreme sensitivity to the economic and psychological manifestations characterizing the various stages of each cycle. During a period of economic growth, for example, consumer confidence is boosted and people intensify their travel activity. In turn, airlines, hotels, and related tourism businesses prosper. Development forges ahead; new businesses and jobs are created. By contrast, during a bust the opposite occurs. As consumer confidence declines, so does demand. Financial instability of tourism firms intensifies, losses occur and in severe cases, a sector-by-sector industry shake-out is initiated. While a fevered cycle of growth and decline can actually be healthy for an economy, since robust churning purges marginal businesses, recovery is often slow and awkward.

The market process also has to be re-adjusted to account for turbulent market dynamics. These dynamics are usually manifested by new technologies allowing the provision of tailored customer service, and increasing the dimensions of connectivity, eventually transforming entire industries. Moreover, such dynamics reflect changing consumer tastes (e.g. volatility of choice that alters the way products and services perform or are sold), and also demonstrate expanding geographical market frontiers (e.g. Asia, Eastern Europe and Latin America that represent and are redefining new markets and competitors). The phenomenal speed of innovation, change and relentless improvement that is occurring throughout the corporate world suggests that tourism's boom and bust syndrome may simply be exacerbated though not dictated by the business cycle. Deregulation, globalization and the technological discontinuity that result from innovations and entrepreneurial changes foster the development of new business models and competencies. In turn companies, and even agglomerations of companies in various destinations, are forced to re-position themselves for survival and future prosperity.

The ability to orchestrate effectively such a transformation in the travel and tourism industry demands an understanding of the major forces driving the cycles of growth and decline. The intent of this chapter, therefore, is to provide a reference point for the overall economic and business activity which affects tourism. This ambitious objective is partially achieved by reviewing and assessing some of the limitations of the prevailing, tourist area, life-cycle model by discussing various aspects of tourism's economic cycles with a particular focus on the lodging sector, and by noting how innovative activity leads to further discontinuity.

Tourist area life-cycle model

The predominant notion that resort areas proceed through cycles of evolution and devaluation has not gone unnoticed. E. Gilbert (1939) and Christaller (1963) were early advocates of the idea that resort areas evolve through stages of discovery, growth, and decline. In 1980 Butler developed a comprehensive model of long-term change that juxtaposed rates and types of market acceptance (demand patterns) with rates and types of development (supply patterns). With explicit concern expressed regarding the finite and possibly non-renewable tourism resource base, and the distinct possibility of destroying the essence and attractiveness of a tourist area, his model portrayed six stages of a 'birth-to-death' cycle (see Figure 13.1 in Chapter 13 of this book).

According to Butler, during the exploration stage, a few visitors discover the charm of a destination, but lack of access, infrastructure and knowledge about the area limit visitation. Local interest in expanding tourism opportunities and the establishment of small-scale enterprises characterize the involvement stage. Local entrepreneurs provide basic services and amenities but marketing efforts are constrained due to costs. A development stage kicks-in if growth potential is evident. Once capital investment infrastructure is obtained, major international tourism suppliers (e.g. accommodation, transportation, and tour wholesaling firms) are courted and vie for a piece of the action. (Butler noted that during this stage control of segments of the industry can be passed from locals to outsiders.) As marketing activities of tourism firms and destinations intensify, the numbers of arriving visitors increase to fill the available capacity. If market potential appears to be great, further development is pursued even though some stakeholder groups may oppose this. Poor planning during this growth stage, however, tends to hasten the arrival of subsequent stages; overcrowding, deterioration of tourism resources, over-development, loss of attractiveness, crass commercialism, and other concerns, including a lack of industry legitimacy within the community, can precipitate decline.

As the visitation growth rates decline and the tourism plant ages and complacency sets in, various sectors of the industry normally consolidate through mergers and acquisitions. Concurrently, marginal businesses go into receivership or declare bankruptcy. Problems such as tourism resource capacity levels being breached or exceeded, and declining levels of visitor satisfaction, eventually lead to stagnation. The destination and its tourism businesses experience extreme difficulty in generating repeat patronage and new markets. Massive re-investment may be required to revive the industry. An effective and well-managed turnaround, however, could result in rejuvenation. The contrary, that is the inability to deal with problems coupled with inefficient operations or strategies that no longer fit the market and competition, will eventually precipitate decline.

While prior case studies generally support Butler's version of the tourist area life cycle as a useful framework for describing, explaining, and analysing the growth and demise of destinations and their markets (Cooper and Jackson 1989; Choy 1992; Ioannides 1992), the model's use as a prescriptive planning, marketing, and forecasting tool has not been fulfilled (Haywood 1986; Cooper and Jackson 1989; Cooper 1994). The reasons are numerous: the definition and delineation of a tourist area and the type of operating business to be included are often unclear, as is an understanding of what constitutes tourism markets. In other words, the heterogeneity and obscurity of some markets make them hard to identify and understand.

A further difficulty, verified by a number of researchers (Haywood 1986; Ioannides 1992), arises from the recognition that different destinations experience varying life-cycle patterns and may miss a stage. Indeed, it can be argued that tourism within a destination may embrace several sectors and different classes of non-competitive products, each of which displays its own unique pattern of evolution. Then there are the conundrums associated with what constitutes 'capacity limits' or overbuilding. Moreover, the determination of appropriate units of measurement has also been proven to be confusing. For example, is the number of arrivals the most appropriate measure for determining growth? Consideration is not often given to such ameliorating variables as the length of stay or the dispersion of visitors within and throughout the tourist area. Neither does the model account for characteristics of tourists or the time of the year in which visits occur.

The inherent problems of the tourist area life cycle make it difficult to find empirical support for the typically generalized descriptions and prescriptions associated with each stage. Key assumptions may be flawed and important dimensions of evolutionary process are overlooked. For instance, one might expect the life-cycle model to have the capacity to reflect the conditions that trigger competitive shake-outs; yet a recurring risk in high-growth markets is that too many competitors enter destinations with unrealistic market share expectations (Aaker and Day 1986). The life-cycle model does not recognize that the strategic window of opportunity (market entry or expansion) opens at different times for different types of business, often a function of the business and real estate cycles (Abell 1978), and that the risks and rewards of entry depend on the choice of timing. Furthermore, the model ignores the fact that supply-side factors can accelerate or decelerate the rate of growth; for example, lethargic business activity in a destination could be an open invitation in other destinations to segment offerings, to be innovative or to promote extensively, thereby accelerating the rate of decline. Because the life-cycle framework assumes no uncertainty, consideration is not given to the trade offs (e.g., maintaining flexibility) involved in confronting uncertainty (Wernerfelt and Karnami 1987). Thus, during the early stages of growth, uncertainty about rates of visitation, the eventual size of the market, or the actions of competitive destinations make decisions about the best destination 'product-service' mix quite complex. There is a tendency, therefore, to ensure that a destination has broad or mass appeal.

In sum, these and other criticisms of the tourist area life cycle imply that there is a necessity to step back and gain a better understanding of the factors that shape the various life-cycle patterns. Marketing researchers studying the product life cycle came to this conclusion years ago (Rink and Swan 1979), as have a few tourism researchers (Haywood 1986; Goodall 1992). In essence, the evolution of tourist areas and product markets reflects the outcome of numerous market, technological and competitive forces that act in tandem with other factors to facilitate or inhibit the rate of growth or decline (Porter 1980). These forces can be categorized within tourism demand and supply systems and the operating tourism environment. With the emphasis of this chapter on the impact of business and economic cycles, however, only a few forces are identified.

Demand system

The dynamics of a destination market are distinctly altered by demographic (Foot and Stoffman 1996) and economic trends, as well as by the evolution of complementary markets (e.g., for sports, recreation, leisure, educational and cultural pursuits). Potential tourism or visitor demand is affected by the comparative advantages of alternative destinations, perceived risks, and barriers to adoption and access (financial, informational, availability). These determinants are dynamic with considerable supply-side influence from: marketing mix variables (price, advertising, personal selling); strategic variables (competitive advantage investment, intensity, resource commitments and allocation patterns); public policy variables (taxation, destination marketing allocations and support standardization); competitive variables (entries and exits of firms, the characteristics, resource and strategies of competition); and the economic variables (changes in interest rates, employment, savings rates, consumption activity).

Supply systems

The rate at which a destination grows or declines is influenced by the number, type, and timing of tourism suppliers entering the market as well as by their particular strategic choices. Investment activity varies according to expectations about market potential and aspirations regarding their own competitive position in the existing and potential markets. Because tourism requires a bundling of different amenities, plus attractions and services, the extent and pattern of tourism development is influenced by the size and types of enterprise that enter all sectors of the industry, the speed of their arrival, their desire to connect with complementary businesses, and the amount and allocation of investment. The presence and behaviour of competitive destinations, particularly their offensive and defensive strategic actions, also need to be taken into consideration.

Operating tourism environment

The dynamics inherent in demand and supply are governed by a continual series of economic, technological, cultural, and political trends and events:

- new technologies that are accelerating the rate at which tourism products and processes must be refined (Slyworthy 1996);
- the availability and stability of costs of resource inputs (food, labour, construction materials) that ultimately determine travel costs and thereby destination choice by visitors;
- availability and changes to destination infrastructure that foster or impede tourism development;
- tourism policies that support or hinder growth of a tourist industry;
- shifts in industry structural configurations that alter the nature of competition (Debbage 1990);
- the economic climate that governs travel patterns and behaviour, pricing arrangement, strategic choices, policy options and so on.

It has been suggested that if the tourism life cycle is to become a framework for decision-makers to think and act more strategically, the model's conceptual validity must be drastically improved (Gardner 1987). Elaboration of the operational context in which tourism products, organizations and destinations exist is one way of moving the life-cycle concept towards a framework that incorporates the dynamics of competitive behaviour in evolving market structures.

Tourism's life cycles and broader economic cycles

Research into the life cycles of destinations dwells on the ebb and flow of supply and demand. However, the literature on this topic has virtually ignored existing statistical data on regional and industry sensitivities to national and international recessions and expansions. This has been largely due to definitional and measurement problems (see Chapters 2–4 in this book). The need, however, to link economic indicators with industry and destination cycles is becoming more pressing. The monitoring, examination, and forecasting of these economic cycles provide the opportunity: to explore timing relationships between the economy and tourism, tourism and other industries, or segments of the tourist industry; to identify and understand how to use the forces within different stages of the business cycle; to develop insight into industry turning points (knowing when to change); and, to mount appropriate management and marketing strategies. Even government policymakers are beginning to find industry indicators useful for fiscal deliberations such as industry-specific tax policies.

In response to this need, Canada has introduced a tourism satellite account (Lapierre and Hayes 1994; see Chapter 3 in this book) and a set of national tourism indicators (Beaulieu-Caron 1997). Developed by Statistics Canada in collaboration with the Canadian Tourism Commission, these indicators have been generated to update the tourism satellite account (TSA) on a quarterly and annual basis. The intention was to provide timely information for monitoring and analysing tourism and related activities. The indicators measure trends from 1986 to date for most components of the TSA, using time series data that match the concept and definition of each TSA component as closely as possible. The indicators cover the domestic supply of tourism commodities, the demand for these commodities by domestic and foreign visitors, and the employment generated within the tourism sectors as a result of the demand.

Preliminary work on comparisons of the trend and cyclical components for each national tourism indicator with the overall Canadian economy is quite revealing. Over the ten-year period 1985–95, it is evident that tourism was a leading industrial sector. The growth in supply of tourism products and services exceeded that of the country's gross domestic product, although there was considerable variation from one tourism commodity to the next. This led to a similar increase in employment that surpassed the growth rate for all industries combined. Despite the dominance of tourism domestic demand, it was the growth in tourism export demand (incoming international travel) that was largely responsible for such an expansion during this period.

Canada's national tourism indicators have also revealed pronounced cycles, with most of the variation in supply and demand statistically explained by the cyclical variation in the overall Canadian economy (over 90 per cent in the supply of tourism products and services; over 80 per cent in tourism demand).

Explanation of variation in tourism export demand, however, depended on economic conditions in tourism-generating countries, exchange rates governing the value of the Canadian dollar, and a host of other events. What is most intriguing and of vital importance is the comparison of the cyclical variation of the gross national product to the cyclical variation in the supply and demand for all tourism products and services. Over the ten-year period, it was determined that cyclical deviations for tourism were twice as large. In other words, when the economy grows, the tourist industry booms; when it stagnates or goes into a deep funk, tourism enters a state of depression. Tourism and general business cycles during this period were synchronized, though there were lags: the cyclical component for domestic tourism demand lagged two to three quarter-periods behind the cyclical component of the gross domestic product.

Using national tourism indicators to examine 'lead-lag' relationships of tourism products and services with the business cycle provides considerable insight into the cyclical aspects of the industry and destinations, and the structure and operation of the industry. Industry forecasts for tourism, using these economic indicators, will be particularly important in examining demand (pleasure and business trips, expenditures and receipts for various sectors), supply, and employment issues. Of course, projection of international travel patterns as they relate to the number of visits, expenditures and so on will necessitate examination of regional and national indicators from the travel-generating countries. It can be anticipated that the annual ritual of economic forecast presentations will indicate how economic climate, consumer confidence levels, employment levels, GDP indices, and inflation contribute to leisure, convention and business travel expenditures and vice versa.

Economic cycles in the lodging industry

The national tourism indicators for Canada identify considerable variation in the cyclical characteristics for different tourism products and services. Therefore, it is vital that more effort be put into developing a better understanding of the cyclical characteristics of each travel-related sector. Recent studies of the lodging sector reveal that room demand and average daily rate correlate positively to the GDP and employment growth. In the United States, for example, from 1989 to 1991 the recession resulted in less disposable income and, consequently, fewer resources for travel. As a result, less travel activity yielded a decline in demand for hotel room-nights. Subsequent to 1992, however, strong job growth, combined with an increasingly healthy national economy, propelled room demand and average rates (Bankers Research Trust 1996).

In the US market, beginning in the late 1980s and continuing through to 1994, growth in average daily rates in the lodging industry did not keep pace with changes in the consumer price index (CPI) due to rampant growth in the supply of rooms. (The CPI is a measure of inflation and represents a key statistic to which earnings are compared in order to assess financial gain.) For the operator or investor, this meant that during this period the annual increases in room rates were not exceeding the annual increases in cost of goods and services. However, 1994 marked the first year in which the increase in room rates outpaced inflation. This trend continued into the next two years. The ability of the industry to raise room rates above the inflation rate was one of the factors contributing to the sector's profitable turnaround. According to Bankers Research Trust (1996), the industry's performance was propelled by the following factors:

- an economic expansion that began in 1992;
- a compound annual rate of demand for hotel rooms of 3.2 per cent, in comparison to a historically low increase to room supply of 1.4 per cent;
- an excess demand averaging 7.5 million room-nights per year between 1991 and 1995, which resulted in rising occupancy rates to an average of 66 per cent in 1996 – the highest level since 1982;
- an increase in average daily rates (6.6 per cent) for the first nine months of 1996 – more than twice the rate of inflation);
- REVPAR (revenue per available room) gains in excess of 7 per cent in 1996, the highest rate since 1980.

The strong performance of the lodging sector during a period of strong economic growth, however, led to a dramatic comeback in new construction. As a result, occupancy was expected to level off in subsequent years with the risk profile of hotel investment intensifying once more.

It would appear however that lodging cycles are not completely in synchronization with economic cycles (Long 1940). They are driven by periods of imbalance between the growth in room supply and room demand. Over the past twenty-five years, industry downturns were typically preceded by at least three years of 3–4 per cent increases in room supply. Two industry downturns were particularly brutal, one during 1974–5 and the other 1990–1. Because lodging is invariably linked with the commercial property market, during the recession of the early 1990s numerous hotel developers and investors saw their fortunes demolished faster than their empty guest rooms.

This cycle of boom and bust has several implications. First of all, removals of ageing inventory from supply seem to occur at a slow rate. Second, changes in average room rates do little to eliminate short-term surpluses or deficits of space: occupancy rates are 'sticky'. Third, developers seem to persist in starting new construction projects long after average room rates have started rising, with some of these reaching the market only after demand has already slowed. This, in turn, creates a glut that can take years to correct.

Explanations concerning the volatility of property cycles have tended to ignore the stickiness of occupancy rates, focusing instead on overbuilding. One popular notion blames myopic and greedy developers who ignore the warning signs. Another notion blames financial incentives offered by government (tax credits) or by lenders (non-recourse loans whereby, if the developer cannot repay, the property is forfeited and the developer is not liable for further payments). These suggestions assume that everyone connected with the industry is mysteriously ignorant, or that they do not take into account the facts.

Grenadier (1995) provides an improved explanation that could be applied to the lodging sector. Filling a low-occupancy hotel by dropping prices may be a sign of lowering quality and, if break-even points are to be maintained, requires boosting occupancy (virtually an impossible task in a cut-throat, low-demand market), which increases marketing costs. Allowing a hotel to become vacant is usually not an effective option; it is costly and may result in expensive revival costs, especially replacement of key staff. Another explanation for the stickiness of room rates is the tendency to wait for average daily rates to rise to a level where short-term benefits exceed the costs of obtaining and looking after new guests.

Overbuilding may also be a rational event. As previously mentioned, there may be occasions to replace old properties (though the speed of removal may be slow); there may be a marked divergence in performance in certain price categories or lodging segments; and development activity in preferred or 'premium' brands may be economically justified even if initial occupancy rates are slow to take off. This is a situation Grenadier compares to a call option on a share, which gives the option's owner the right to buy that share at a specified 'strike' price in the future. Because share prices are volatile, this option is valuable even if the current share price is less than what was originally paid. If the price jumps above the price paid, the holder can exercise it and pocket the difference; if not, he loses little. The more volatile the price of the underlying share, and the longer the option's period of validity, the more valuable the option will be.

These factors, in conjunction with project completion periods that take two or three years, make the industry more prone to overbuilding and may explain why property cycles are so protracted. The structural changes brought about by the extensive overbuilding and over-leveraging of lodging properties during the 1980s and the broader participation of capital markets in the lodging industry, however, may bring a new measure of discipline and stability. Nevertheless, economic weakness or recession does cause a decline in occupancy, a weakening of pricing power and slow growth for the industry, particularly in some destinations.

It is worth noting at this juncture that the pervasiveness of cyclical fluctuations in all property sectors has geographers as well as urban and regional planners speculating on strong cyclical tendencies. Each successive boom is associated with a major wave of development or redevelopment, and each slump with a period of stagnation and urban area decline (Barras 1987). An understanding of the incidence of building cycles, therefore, may be essential for interpreting tourist area life cycles. Moreover, such an understanding contributes to the debate about destination crises as manifested by the dereliction of certain areas and the collapse of obsolete infrastructure.

Innovation and discontinuity

Along with the inevitable short swings associated with business cycles, there are swings within tourism, as in many industries, that economists such as Kondratiev (1922) have associated with long waves of economic development. Kondratiev observed that during recessions important discoveries and inventions associated with production and communications are made and applied on a large scale. Schumpeter (1963) also observed that market dynamics are brought about by technological change and that competitive advantage grows out of improvement, innovation, and change which contemporary theorists say can only be sustained through the relentless search for value (Rumelt 1984 and 1987; Porter 1990; Slyworthy 1996).

Schumpeter's vision of competition as 'creative destruction' coincides with the current view that strategy formulation is centering more on entrepreneurial discovery, that is, the reformulation of a product's function, the development of new business processes, new distribution and communication channels, or the discovery of dimensions of competition that other businesses have overlooked (Gluck, *et al.* 1980). It should be pointed out, however, that as knowledge changes, markets continually change. In other words, continuous changes in the states of knowledge produce new disequilibrium situations and, therefore, new profit opportunities (Conner 1991) which eventually decay as successful practices and market conditions change (W. Beaver 1970).

Today, this long wave of economic expansion and entrepreneurial discovery is unprecedented in size and scope. A vast expansion of economic freedom and property rights, coupled with reductions in the scope of government and an explosion in trade and private investment, has produced, according to the International Monetary Fund (1997), a world trade growth that in the past few years has nearly doubled that of the prior two decades. Multinational hotel and tourism corporations are in the forefront of the globalization process and among its biggest beneficiaries. Aided by domestic deregulation, flexible labour markets, and their own openness to new technology, many firms are transforming themselves through compelling innovation imperatives.

As a global industry, tourism is breeding diversity in the market-place, innumerable competitors, more choices for the traveller, and a wealth of opportunities. Not only has high technology introduced speed and knowledge as a basis for competition, but it has pressured organizations to dismantle processes that were once seen as effective. Contributing further to the turmoil, tourism's critical constituencies – visitors, shareholders, communities, managers and employees – are simultaneously more informed and demanding. Innovative energy is apparent in every domain of tourism. Significant technology and product developments and improvements are being introduced on a daily basis – business software packages, Internet sites, electronically equipped guest rooms, property management systems. Processes are being altered as well. Phases of the hotel's guest cycle can now be altered or circumvented through self-registration and check-out systems.

Technology breakthroughs are also cyclical. In the past, technology development in the industry appeared to run in seven-to-nine-year cycles. This is no longer the case. Recycling appears to be occurring in pure technology every other year. For the tourist industry, the technology development cycle probably breaks into three-to-five-year intervals. According to industry insiders, the next big cycle will focus on the fully integrated system that brings customer, statistical and financial information together through a combination of property management, central reservations, data warehouse and distributed database systems. Most major international lodging firms (e.g. Marriott, Hilton, ITT Sheraton, Holiday Inn) have dedicated technology departments that are on the forefront of this 'knowledge-value' revolution.

Paralleling the compression of product and technology life cycles is the rapid change in organizations themselves. The hotel business is evolving from a business concentrating on physical assets to one centering on virtual assets. Brands and their distribution systems are becoming an increasingly important competitive advantage. As lodging companies divest themselves of real estate holdings and concentrate on management, the importance of the brand intensifies. Indeed, as competition increases nationally and globally, smaller chains and independent hotels face a strategic decision – either stand alone or band together for their collective good. Today, a wave of consolidation – mergers and acquisitions – is taking place. Companies are looking for strategic partnerships that help extend their market share and achieve economies of scale and synergy. With every new deal come divestitures and stronger Goliaths (with access to low-cost capital) with which other companies must compete in the market-place. To meet long-term growth objectives, and in response to consumer needs, there has been significant activity in the introduction of new hotel brands as well as segmenting and restructuring existing brands.

Conclusion

Examination of economic cycles and innovation activity in the context of tourism, or more specifically in geographically bounded areas (tourist regions), points to the constancy and regularity of changing operating environments and the need to adapt. Indeed, this is a dynamic process based on competition within commercial product-markets, not simply competition for a scarce and non-renewable tourism resource base. However, the value of the existing tourist area life cycle is limited in defining the factors shaping life-cycle patterns, and explaining the adaptions of tourism businesses and destinations during the evolution of product markets and tourism regions. By ignoring the external operating environment and the way in which this affects the industry, the usefulness of the life-cycle model as both a descriptive and prescriptive tool is undermined.

On the basis of preliminary empirical data (Canada's national tourism indicators), the impacts of business and other economic cycles on tourism appear serious and, indeed, catastrophic in extreme cases. Use of this data, however, creates new possibilities associated with analysing the economic structure of tourism. For example, it will be possible to determine more precisely how tourism demand is affected by the level of disposable income, the price of commodities, foreign exchange rates, deregulation, and marketing expenditures. This new knowledge could lead to the development of government policies and business strategies that could avert decline, and foster and generate greater tourism activity.

With evidence that economic and business cycles have profound impacts on individual businesses and destinations, it is vital that the tourist area life-cycle model be expanded to include a supply-side theory of 'product-market' evolution that would account for changes in the operating and competitive environment over time. For example, it is evident that in any given destination, changing economic circumstances and crises precipitated by business and economic cycles, rapid technological change, and collapsing product and organizational life cycles, alter competitive profiles, positions, and resources of competing large and small businesses in different ways. A supply-side approach could also help in developing an understanding of the differences between established and new firms at various stages of market evolution, particularly as related to the timing of market entry and exit. As such, it may be possible to develop more appropriate typologies of strategies for destinations and businesses in these destinations.

Part F

SYNTHESIS AND NEW DIRECTIONS

16

CONCLUSION: THE COMMODIFICATION OF TOURISM

Keith G. Debbage and Dimitri Ioannides

The machinery of tourist production

We began this book by calling for a better integration of tourism's supply-side within the spatial context of the tourism geography and economic geography literature. Unfortunately, geographers have thus far been reluctant to embrace the processes of commodification that have transformed tourism experiences into more marketable commodities. Yet, it is the infrastructure of production and the geographically uneven accumulation of capital that may play the most powerful role in manipulating consumption levels and behavior patterns. Although changes in consumer demand and the evolution of increasingly more sophisticated consumer preferences can play substantive roles in shaping the tourist product, it is the actual 'machinery of production' (e.g. tour operators, travel agents, airlines, and hotels) that helps to manipulate and facilitate origin–destination tourist flows across the world.

Consider, for example, the complexity of an international trip for a middle-aged adult in contemporary Manhattan. The initial decision to travel to the United Kingdom was triggered by a generic print advert in *The New York Times* that was paid for jointly by the British Tourist Office and British Airways – an example of the increasing levels of collaboration and rationalization of costs that is now commonplace between functionally related travel suppliers, especially when marketing the end product.

Our individual acquires from a local 'mom and pop' travel agent a glossy brochure prepared by a New York-based specialist tour operator, which niche-markets heritage-based packaged vacations to northern England. Additional information is provided via the Internet by various regional destination marketing organizations based in northern England that our individual accesses via the computer at his or her apartment in downtown Manhattan. The individual eventually books the resort, airline, and sightseeing tours through a California-based travel agent that has a strategic alliance agreement with the New York tour operator. The tour operator does not have direct contact with the consumer. Instead, the tour operator focuses solely on the core business of negotiating packaged deals with key suppliers and providing the necessary glossy brochures to a network of travel agents that 'drum-up' business for the tour operator on commission. By doing so, the tour operator can focus on the core business of providing competitively priced packaged vacations in a sector of the economy that traditionally operates on wafer-thin profit margins.

The air carrier, in the meantime, is a scheduled airline – not a chartered airline – that flies out of Newark, New Jersey. Nevertheless, the carrier does allow tour operators to block-book seats in advance. It was the lower fares which attracted the New York tour operator to the carrier when putting together the overall travel package. The carrier flies out of Newark because of the lower landing fees relative to JFK Airport in New York. The air carrier leases the aircraft at a competitive rate from a leasing agency based in Dublin, Ireland, that specializes in aircraft operations. Incidentally, the carrier has also out-sourced its entire maintenance operations in order to reduce labor costs. The air carrier also has a code-sharing agreement with a British commuter airline so that 'seamless service' (e.g. one-stop baggage drop-off and pick-up) is provided when changing planes at Gatwick Airport in London. The commuter carrier operates a propeller-based shuttle service with frequent departures to the historic city of York in northern England – the final destination. The agreement negotiated between the major jet carrier and the smaller commuter airline allows for a synergy and economies of scale. This ensures that the commuter airline draws significant levels of feeder traffic which enable it to offer competitive fares relative to its competitors. The limited number of seats on any given commuter flight ensures a high yield and an appropriate matching of demand with supply.

However, without the sophisticated information technology base and computer reservation system (CRS) developed by the major jet carrier (which has the research and development monies to do such a thing), the complex shuttling of passengers from one carrier to the other would have been impossible. In fact, the entire trip was reserved by the travel agent in California using the major carrier's CRS – the travel agent received a 10 per cent commission for routing traffic through the host carrier's CRS.

Upon arrival at the destination, our individual is transported to the hotel by an airport shuttle service. The competitive hotel room rates were made possible because the tour operator bulk-purchased a significant number of rooms in advance. The New York tour operator also has a small financial interest in the American-based hotel chain that operates this hotel property in England, and the hotel chain participates in the joint frequent-flyer program offered by the jet carrier that transported our individual from Newark to London. Although the hotel is part of a large chain operation, the holding company has niche-marketed its six subsidiaries by developing a sophis-ticated, brand super-segmentation strategy such that the consumer perceives the holding company to operate six functionally different hotel chains. Our individual stays at the full-service, de-luxe subsidiary that focuses on downtown locations within central business districts, and is convinced that the hotel provides a high-quality service unlike those anonymous, sterile large-scale chains!

Although the trip is primarily for pleasure, our individual has arranged a business meeting with a client in one of the conference meeting rooms provided by the hotel. Our individual is a software engineer for a large telecommunications firm in Manhattan, and the trip has provided the opportunity to negotiate a deal with a functionally-related British firm which has its corporate headquarters in Geneva. It is at this point that the trip changes from a consumer service-based activity focused on final demand to a producer service-based activity focused on intermediate demand.

Our individual has decided to expand his or her geographic horizons by purchasing a rental car at the destination, although this decision was made earlier when booking the overall trip in New York. The New York tour operator was able to offer a competitively priced fly/drive package that was negotiated when establishing the overall packaged-vacation product. The chosen rental car company is also tied to the frequent flier program offered by the jet carrier and hotel chain, and this generates a certain level of brand loyalty in our individual. While touring the heritage attractions of northern England by car, our individual begins to appreciate the diversity of the tourist product. The tourist experience includes visiting local and chain restaurants, dance clubs, museums and stately homes, craft centers, and various retail and souvenir stores, while also participating in more intangible non-market-based activities such as viewing landscapes and scenery, hiking, and art appreciation. Our individual's tourist expenditures begin to ripple through the regional economy as he or she purchases more indirect tourist-related goods and services such as gasoline for the rental car, and foodstuffs from the local supermarket, while also making an unplanned visit to a medical clinic. It is at this point when it becomes difficult to determine if this form of expenditure can be categorized as a basic economic activity or a non-basic economic activity in the context of economic base theory.

Our individual also discovers the merit of purchasing the regional tourist attractions coupon booklet put together by the regional marketing agency, for it offers numerous discounts on a wide range of tourist attractions, but only when taken in combination with each other. Even more importantly in geographic terms, the quasi-public regional tourist board has provided a scenic tour map that effectively 'steers' visitors to the appropriate locations. To make matters even more complex, our individual plans to visit an entirely different destination and utilize an entirely different set of tourism commodities on his or her next trip.

This crude, but hopefully useful, illustration demonstrates the very complex and powerful 'machinery' that acts to facilitate the 'tourist gaze'.

While much has been written about the social relations of the tourist gaze, considerably less attention has been focused on the infrastructure of production that both manipulates and permits the tourist experience actually to happen. Although the travel and tourism industry has manipulated and fostered origin–destination tourist flows since before the days of Thomas Cook, what is new today is the increasingly sophisticated nature of the apparatus. The restructuring of production in the tourist industry has recently been supplemented by:

- the rise of networks of flexible specialization (Ioannides and Debbage, Chapter 6 in this book; Mackun, Chapter 14);
- advances in telecommunications and information technology (Milne and Gill, Chapter 7);
- a form of industrial restructuring that favors processes of industrial concentration, economies of scale and scope, and the establishment of large-scale alliance networks (Ioannides, Chapter 8; Wheatcroft, Chapter 9; Milne and Pohlmann, Chapter 10);
- the commodification of place and the emergence of cities as places of consumption (Hall, Chapter 11; Beauregard, Chapter 12);
- the emergence of more sophisticated forms of consumer demand (Uyssal, Chapter 5).

The reassertion of the distinctiveness of the locale as a place to invest in is crucial to all these processes, particularly as the consumer becomes more discerning when choosing a destination. But the investment capital needed to create and embellish these same destinations is increasingly tied up in international circuits of finance involving some of the most powerful business corporations in the world (Haywood, Chapter 15).

New research opportunities

Although the contributions to this edited collection reflect the diversity of the paradigms and interpretations of the factors driving change in the tourist industry, we shall conclude by attempting to identify some of the key themes, spatial consequences and new research directions of an economic geography of the tourist industry.

First, we argue that without a rigorous conceptualization of what comprises the tourist industry, the economic geography of tourism literature will continue to languish in the intellectual backwaters of the discipline (Debbage and Daniels, Chapter 2). The development of a tourism satellite account (Smith, Chapter 3) and the introduction of the North American industrial classification system (NAICS) with more refined travel-related industrial definitions (Roehl, Chapter 4) offer tremendous opportunities for industrial geographers looking for more detailed empirical data on the tourist industry.

Second, we believe that a more rigorous analysis of how post-Fordism and flexible specialization practices have affected the travel and tourism industry could enhance our understanding of post-Fordism's impact on the contemporary economy. The travel industry has a well-established, historical tradition of employing mainly numerical, but also functionally flexibile strategies primarily because of the volatility and seasonality of the industry. Furthermore, even the larger travel-related corporations are now emphasizing flexibility through brand super-segmentation strategies and the more sophisticated application of information technologies. Whereas the inherent fact that the increased levels of industrial concentration seem to contradict the emphasis on sophisticated niche-marketing and the uniqueness of locality, it also seems to suggest a neo-Fordist approach that emphasizes a polyglot of production philosophies (Ioannides and Debbage, Chapter 6).

Third, the introduction of information technologies and CRSs may allow for more flexible, alternative forms of tourism, but others question the uncritical embrace of CRSs in facilitating such a trend (Milne and Gill, Chapter 7). The high cost of information technologies may work against the development of inter-firm networks and strategic alliances, particularly among smaller travel firms and hotels, while tour operators seem particularly resistant to some forms of information technology. (Ironically enough, this resistance is partly due to the inordinate complexity and flexibility of many packaged vacation products.) (Ioannides, Chapter 8.)

Nevertheless, the introduction of telecommunications technologies in tourism can only accentuate the continued spatial division of labor by front and back-office functions, while also encouraging higher levels of productivity. For example, the tourist industry shows signs of entering a period of 'labor-shedding' as the application of information technologies allows practitioners to do such things as automate check-in and check-out, and by-pass labor-intensive activities such as travel agents by providing travel information directly to the consumer through the Internet. Indeed, not all sectors of the travel and tourism industry are technology laggards, and the importance of entrepreneurship, innovation, and trust and reciprocity, in the context of tourism labor markets, are examined by several of our contributors (Milne and Pohlmann, Chapter 10; Shaw and Williams, Chapter 13; Mackun, Chapter 14).

Fourth, the continued commodification of culture and place that has induced the 'place-wars' and 'place-marketing' of the 1990s, has led to the rapid recycling of space as more and more destinations attempt to compete and respond to the rapidly changing tastes of the consumer (Hall, Chapter 11; Beauregard, Chapter 12). For example, many of America's downtown areas have reconfigured themselves and are now better known as places of consumption than as places of production (e.g. Boston's Faneuil Hall and Baltimore's Inner Harbor) (Plate 16.1).

16.1 Global coca-colanization: the Coca-Cola museum in downtown Atlanta, Georgia

Source: Dimitri Ioannides

Finally, it is vital that any conceptualization of the tourist industry accounts for the highly competitive external environment within which any given supplier operates. The industry has traditionally been highly volatile and evidence exists to suggest that it is extraordinarily sensitive to changes in the macro-economic environment (Haywood, Chapter 15). The volatility of these changes is perhaps best seen in the tour operator industry where the major operators essentially market holiday type such that place is readily substitutable – contradicting the ethos of the uniqueness of locality.

We could continue this agenda at length, but we shall end by arguing that all economic geographers should be interested, in some way, in both the apparently fundamental shift from a focus on work and production to a focus on leisure and consumption in contemporary society, and how these changes have been facilitated by such things as information technologies, flexibility and place commodification. We leave it to future researchers to uncover whether all this industrial restructuring in the tourist industry simply involves new ways of exerting good-old market power or whether it really reflects the incipient stages of the dawning of a new era of more sustainable, alternative forms of tourism. On the evidence of this book, we suspect the former.

BIBLIOGRAPHY

Aaker, D.A. and Day, G.S. (1986) 'The perils of high growth markets', *Strategic Management Journal* 7: 409–21.

Abell, D.F. (1978) 'Strategic windows', *Journal of Marketing* 42: 21–26.

Agarwal, S. (1997) 'The public sector: planning for renewal', in G. Shaw and A.M. Williams (eds) *The Rise and Fall of British Coastal Resorts*, London: Cassell, 137–58.

Air Transport World (1994) 'Gold in the galleys' (August): 89–90.

—— (1997) 'Bilateral ballistics' (February): 53–61.

Allen, J. (1988) 'Service industries: uneven development and uneven knowledge', *Area* 20: 15–22.

—— (1992) 'Services and the UK space economy: regionalization and economic dislocation', *Transactions of the Institute of British Geographers NS* 17: 292–305.

Amin, A. (1989) 'Flexible specialization and small firms in Italy: myths and realities', *Antipode* 21: 13–34.

Amin, A. and Robins, K. (1990) 'Industrial districts and regional development: limits and possibilities', in F. Pyke, G. Becattini, and W. Sengenberger (eds) *Industrial Districts and Inter-Firm Cooperation in Italy*, Geneva: International Institute for Labour Studies, 185–219.

Anderson, J.E. (1984) *Public Policy Making*, 3rd edn, New York: CBS College Publishing.

Apostolopoulos, Y., Leivadi, S. and Yiannakis, A. (1996) *The Sociology of Tourism*, London: Routledge.

Arbel, A. and Ravid, S. A. (1985) 'On recreation demand: a time-series approach', *Managerial and Decision Economics* 17,6: 979–90.

Archdale, G. (1992) *Computer Reservation Systems and Public Tourist Offices*, Madrid: WTO.

Archer, B.H. (1980) 'Forecasting demand – quantitative and intuitive techniques', *Tourism Management* 1,1: 5–12.

—— (1995) 'Importance of tourism for the economy of Bermuda', *Annals of Tourism Research* 22,4: 918–30.

Armstrong, C.W.G. (1972) 'International tourism: coming or going: the methodological problems of forecasting', *Futures* 4,2: 115–25.

Arnaut, G. (1994) 'PC-based systems ease do-it-yourself booking', *The Globe and Mail* 22 February: C1.

Ashworth, G.J. (1992) 'Is there an urban tourism?' *Tourism Recreation Research* 17: 3–8.

Association of District Councils (1993) *Making the Most of the Coast*, London: Association of District Councils.

Aviation Week and Space Technology (1993) 'Competition keener in airline upkeep market' (May): 40–41.

—— (1996a) 'Homework: is the traditional airline reservation center on the endangered species list?' (June): 138–40.

—— (1996b) 'Trying to raise the crossbar' (June): 23 .

—— (1996c) 'Cost pressure driving MRO-airline cooperation' (18 March): 35.

—— (1997) 'Competition floodgates opened in Europe' (7 April): 32–33.

Bachrach, P. and Baratz, M.S. (1970) *Power and Poverty, Theory and Practice*, New York: Oxford University Press.

Bagguley, P. (1990) 'Gender and labour flexibility in hotel and catering', *The Service Industries Journal* 104: 737–47.

Bagnasco, A. (1977) *Tre Italie: la problematica territoriale dello sviluppo italiano*, Bologna: Il Mulino.

Balfet, M. (1993) 'Les RGDS: une revolution mondiale de la distribution touristique', *Espaces* 12 (May/June): 20–31.

Bankers Research Trust (1996) *US Lodging Industry Overview*, New York: BT Securities Corporation.

Barras, R. (1987) 'Technical change and the urban development cycle', *Urban Studies* 24: 5–30.

Bartlett, C.A. and Ghosal, S. (1989) *Managing Across Borders – The Transnational Solution*, Boston: Harvard Business School Press.

Beaulieu-Caron, L. (1997) 'National tourism indicators. A new tool for analyzing tourism in Canada', *Travel-log Statistics Canada* cat. no. 87-003-XPB, Winter: 1–6.

Beauregard, R.A. (1989) 'Postwar spatial transformations', in R. A. Beauregard (ed.) *Atop the Urban Hierarchy*, Totowa, NJ: Rowman & Littlefield, 1–44.

—— (1993a) 'Industrial diversification as economic policy', in D. Fasenfast (ed.) *Local Economic Development Policy Formation*, London: Macmillan.

—— (1993b) *Voices of Decline: The Postwar Fate of US Cities*, Oxford: Basil Blackwell.

—— (1993c) 'Constituting economic development', in R.D. Bingham and R. Mier (eds) *Theories of Local Economic Development*, Newbury Park, CA: Sage Publications, 267– 83.

Beaver, A. (1992) 'Hotel CRS: an overview', *Tourism Management* 13: 15–21.

Beaver, W.H. (1970) 'The time series behaviour of earnings', *Journal of Accounting Research* 8: 62–69.

Begg, I. (1993) 'The service sector in regional development', *Regional Studies* 27,8: 817–25.

Belisle, J.F. (1983) 'Tourism and food production in the Caribbean', *Annals of Tourism Research* 10,3: 497–513.

—— (1984) 'The significance and structure of hotel food supply in Jamaica', *Caribbean Geography* 1,4: 219–33.

Bell, D. (1973) *The Coming of Post-Industrial Society: A Venture in Social Forecasting*, New York: Basic Books.

Bennett, M. (1993) 'Information technology and the travel agency: a customer service perspective', *Tourism Management* 14,4: 259–66.

Bennett, M. and Radburn, M. (1991) 'Information technology in tourism: the impact on the industry and supply of holidays', in M.T. Sinclair and M.J.Stabler (eds) *The Tourism Industry: An International Analysis*, Oxon: C.A.B. International, 45–65.

Berry, B., Conkling, E. and Ray, M. (1993) *The Global Economy: Resource Use, Locational Choices, and Regional Specialization in the Global Economy*, Englewood Cliffs, NJ: Prentice Hall.

Berson, L.E. (1982) 'Philadelphia: the evolution of economic urban planning, 1945–1980', in J.C. Raines, L.E. Berson and D. McI. Gracies (eds) *Community and Capital in Conflict*, Philadelphia: Temple University Press, 171–97.

Beyers, W.B. (1992) 'Producer Services', *Progress in Human Geography* 16,4: 573–83.

Bianchini, F. (1991) 'The Third Italy: model or myth', *Ekistics* 350: 336–45.

Blank, U. (1989) *The Community Tourism Industry Imperative: The Necessity, the Opportunities, Its Potential*, State College, PA: Venture Publishing.

Bluestone, B. and Harrison, B. (1982) *The Deindustrialization of America*, New York: Basic.

Boettcher, J. (1996) 'NAFTA prompts a new code system for industry – the death of SIC and the birth of NAICS', *Database* 19,2: 42–45.

Bond, M.E. and Ladman, J.R. (1982) 'A strategy for developing tourism', *Research in Tourism* 2,2: 45–61.

Boniface, B.G. and Cooper, C. (1994) *The Geography of Travel and Tourism*, 2nd edn, Oxford: Butterworth-Heinemann.

Boorstein, D.J. (1964) *The Image: a Guide to Pseudo-events in America*, New York: Harper and Row.

Bourdieu, P. (1984) *Distinction: A Social Critique of the Judgement of Taste*, Andover, Hants: Routledge and Kegan Paul.

Bowes, M.M. (1978) 'Spatial competition and the theory of international trade', Ph.D. thesis at University of Illinois at Urbana-Champaign, Michigan: University Microfilms International.

Boyer, C. (1992) 'Cities for sale: merchandising history at South Street Seaport', in M. Sorkin (ed.) *Variations on a Theme Park: The New American City and the End of Public Space*, New York: Hill and Wang, 181–204.

Brandon, P. (1991) *Thomas Cook: 150 Years of Popular Tourism*, London: Secker and Warburg.

Breathnach, P., Henry, M., Drea, S. and O'Flaherty, M. (1994) 'Gender in Irish tourism employment', in V. Kinnaird and D. Hall (eds) *Tourism: A Gender Analysis*, London: John Wiley & Sons, 52–73.

Britton, R.A. (1978) 'International tourism and indigenous development objectives: a study with special reference to the West Indies', Ph.D. dissertation, University of Minneapolis, MN.

—— (1979) 'Some notes on the geography of tourism', *Canadian Geographer* 23,3: 276–82.

—— (1981) 'Some notes on the geography of tourism: a reply', *Canadian Geographer* 25,2: 197–99.

Britton, S.G. (1981) *Tourism, Dependency and Development: A Mode of Analysis*, No. 23, Canberra: Department of Development Studies, Australian National University.

Britton, S.G. (1982) 'The political economy of tourism in the Third World', *Annals of Tourism Research* 9,3: 331–58.

—— (1991) 'Tourism, capital, and place: towards a critical geography of tourism', *Environment and Planning D. Society and Space* 9,4: 451–78.

Brooks, S. (1993) *Public Policy in Canada*, Toronto: McClelland and Stewart.

Brown, B. (1987) 'Recent tourism research in S.E. Dorset', in G. Shaw and A.M. Williams (eds) *Tourism and Development: Overviews and Case Studies of the UK and the SW Region*, Exeter: Department of Geography, University of Exeter.

Brown, B. and Hankinson, A. (1986) *Final Report of the National Small Hotel Study Phase I*, London: ESRC.

Brown, H. (1974) 'The impact of the tourist industries on the agricultural sectors; the competition for resources and food demand aspects', *Proceedings of the 9th West Indies Agricultural Economics Conference*, University of West Indies.

Brownlie, D. (1994) 'Market opportunity analysis: A DIY approach for small tourism enterprises', *Tourism Management* 15,1: 37–45.

Buhalis, D. (1993) 'RICRMS as a strategic tool for small and medium tourism enterprises', *Tourism Management* 14,5: 366–78.

—— (1995) 'Information and telecommunications technology as a strategic tool for economic, social, cultural and environmental benefits enhancement of tourism at destination regions', paper presented at the Indonesian–Swiss Forum on Culture and International Tourism, Yogyakarta, 22–26 August.

Bull, P.J. and Church, A.P. (1994) 'The geography of employment change in the hotel and catering industry of Great Britain in the 1980s: a subregional perspective', *Regional Studies* 28,1: 13–25.

Burkart, A.J. and Medlik, S. (1981) *Tourism: Past, Present, and Future*, London: Heinemann.

Burridge, K. and Milne, S. (1996) 'Tourism, development and travel distribution technologies: the case of the Cook Islands', paper presented to the Pacific Rim Tourism 2000 Conference, Rotorua, 3–5 November.

Butler, R.W. (1980) 'The concept of a tourist area cycle of evolution: implications for management of resources', *The Canadian Geographer* 24: 5–12.

Canadian Tourism Commission (1995) *Technology and Tourism Marketing in Canada*, Ottawa: CTC.

Canedy, D. (1996) 'For Sabre Group, some long-term reservations', *The New York Times* 6 October: F6.

Capecchi, V. (1990) 'A history of flexible specialisation and industrial districts in Emilia-Romagna', in F. Pyke, G. Becattini, and W. Sengenberger (eds) *Industrial Districts and Inter-Firm Cooperation in Italy*, Geneva: International Institute for Labour Studies, 21–36.

Carey, K. (1991) 'Estimation of Caribbean tourism demand: issues in measurement and methodology', *Atlantic Economic Journal* 19,3: 13–18.

Cazes, G. (1972) 'Le role du tourisme dans la croissance économique: reflexions partir de trois examplaires Antillais', *The Tourist Review* 27: 43–47.

Chadwick, R.A. (1981) 'Some notes on the geography of tourism: a comment', *Canadian Geographer* 25,2: 191–97.

Chamberlain, J. (1992) 'On the tourism trail', *North and South* September: 88–98.

Chang, T.C., Milne, S., Fallon, D. and Pohlmann, C. (1996) 'Urban heritage tourism: the global–local nexus', *Annals of Tourism Research* 23: 284–305.

Chapman, K. and Walker, D.F. (1991) *Industrial Location: Principles and Policies*, Oxford: Basil Blackwell.

Chervenak, L. (1992) 'Global CRS: the race continues', *IAHA* June/July: 22–27.

Choy, D.J.L. (1992) 'Life cycle models for Pacific Island destinations', *Journal of Travel Research* 30,3: 26–31.

Christaller, W. (1963) 'Some considerations of tourism location in Europe', *Paper of the Regional Science Association* 12: 95–105.

Christopherson, S. (1989) 'Flexibility in the US service economy and the emerging spatial division of labour', *Transactions of the Institute of British Geographers NS* 14: 131–43.

—— (1993) 'Market rules and territorial outcomes: the case of the United States', *International Journal of Urban and Regional Research* 17: 274–88.

Christopherson, S. and Noyelle, T. (1992) 'The US path toward flexibility and productivity: the re-making of the US labour market in the 1980s', in H. Ernste and V. Meier (eds) *Regional Development and Contemporary Industrial Response: Extending Flexible Specialisation*, London: Belhaven, 163–78.

Clark, C. (1940) *Conditions of Economic Progress*, London: Macmillan.

Clark, Jr., C.E. (1989) 'Ranch-house suburbia: ideals and realities', in L. May (ed.) *Recasting America*, Chicago: University of Chicago Press, 171–91.

Cleverdon, R. (1992) 'Global tourism trends: influences and determinants', in D.E. Hawkins, J.R.B. Ritchie, F. Go and D. Frechtling (eds) *World Travel and Tourism Review Indicators, Trends and Issues* 2, Wallingford: CAB International.

CLIA – Cruise News (Internet) (1997) 'Surprise! You – yes, you – might be in the market for a cruise'. http://www.ten-io.com/clia/news/ (4 May).

Cloke, P. (1992) *Policy and Planning in Thatcher's Britain*, Oxford: Pergamon.

Coffey, W. J. (1992) 'The role of producer services in systems of flexible production', in H. Ernste and V. Meier (eds) *Regional Development and Contemporary Industrial Response: Extending Flexible Specialisation*, London: Belhaven. 133–46.

—— (1995) 'Producer services research in Canada', *Professional Geographer* 47: 74–81.

Coffey, W.J. and Bailly, A.S. (1992) 'Producer services and systems of flexible production', *Urban Studies* 29: 857–68.

Cohen, E. (1995) 'Contemporary tourism,' in R. Butler and D. Pearce (eds) *Change in Tourism*, London: Routledge, 12–29.

Collier, D. (1989) 'Expansion and development of CRS', *Tourism Management* 10,2: 86–88.

—— (1993) 'Information systems in the distribution of tourism products', *Viamericas* no. 2, June, Montreal: WTO.

Conner, K.R. (1991) 'A historical comparison of resources-based theory and five schools of thought within industrial organization economies: Do we have a new theory of the firm?', *Journal of Management* 17,1: 121–54.

Cooke, P. (1983) *Theories of Planning and Spatial Development*, London: Hutchinson.

Cooke, P. and Wells, P. (1989) 'Strategic alliances in computing and communications', paper presented at a symposium on 'Regulation, innovation, and spatial development', University of Wales, 13–15 September.

Cooper, C. (1994) 'The destination life cycle: an update', in A.V. Seaton (ed.) *Tourism: The State of the Art*, New York: John Wiley & Sons, 340–46.

Cooper, C. (1997) 'Parameters and indicators of the decline of the British seaside resort', in G. Shaw and A. Williams (eds) *The Rise and Fall of British Coastal Resorts*, London: Cassell, 79–101.

Cooper, C.P. and Jackson, S. (1989) 'Destination life cycle: The Isle of Man case study', *Annals of Tourism Research* 16,3: 377–98.

Corner, J. and Harvey, S. (1991) *Enterprise and Heritage: Crosscurrents of National Culture*, London: Routledge.

Courault, B. and Romani, C. (1992) 'A re-examination of the Italian model of flexible production from a comparative point of view', in M. Storper and A.J. Scott (eds) *Pathways to Industrialization and Regional Development*, New York: Routledge, 205–15.

Craik, J. (1991) *Resorting to Tourism: Cultural Policies for Tourist Development in Australia*, North Sydney: Allen and Unwin.

Crompton, J.L. (1979) 'Motivations of pleasure vacation', *Annals of Tourism Research* 6,4: 408–24.

Crotts, J.C. and van Raaij, F.W. (eds) (1994) *Economic Psychology of Travel and Tourism*, New York: Haworth Press.

Crouch, G.I. (1992) 'Effect of income and price on international tourism', *Annals of Tourism Research* 19,4: 643–64.

—— (1994a) 'The study of international tourism demand: a survey of practice', *Journal of Travel Research* 32,4: 41–55.

—— (1994b) 'The study of international tourism demand: a review of findings', *Journal of Travel Research* 32,2: 12–23.

Cunningham, J.B. and Lischeron, J. (1991) 'Defining entrepreneurship', *Journal of Small Business Management* 29,1: 45–61.

Curtis, M. (ed.) (1962) *The Nature of Politics*, New York: Avon Books.

Cyberatlas (1996) 'Demographics', http://www.cybratlas.com/demographics.html.

D'Amore, L. (1983) 'Guidelines to planning harmony with the host community', in P.E. Murphy (ed.) *Tourism in Canada: Selected Issues and Options*, Victoria: Department of Geography, University of Victoria.

Daniels, P.W. (1985) *Service Industries: A Geographical Analysis*, London: Methuen.

—— (1986) 'Producer services in the post-industrial space economy', in R. Martin and R. Rowthorn (eds) *The Geography of Deindustrialisation*, London: MacMillan, 291–321.

—— (1991) 'Some perspectives on the geography of services', *Progress in Human Geography* 15: 37–46.

—— (1993) *Service Industries in the World Economy*, Oxford: Blackwell.

Daniels, P.W. and Moulaert, F. (eds) (1991) *The Changing Geography of Advanced Producer Services*, London: Belhaven Press.

Dann, G.M.S. (1996a) *The Language of Tourism: A Sociolinguistic Perspective*, Wallingford: CAB International.

—— (1996b) 'Tourists image of a destination: an alternative analysis', in D.R. Fesenmaier, J.T. O'Leary and M. Uysal (eds) *Recent Advances in Tourism Marketing Research*, New York: The Haworth Press, 41–55.

Davis, G., Wanna, J., Warhurst, J. and Weller, P. (1993) *Public Policy in Australia*, St Leonards: Allen and Unwin.

de Kadt, E. (ed.) (1979) *Tourism: Passport to Development*, Oxford: OUP.

de Souza, A.R. (1990) *A Geography of World Economy*, Columbus: Merrill.

de Souza, A.R. and Stutz, F.P. (1994) *The World Economy: Resources, Location, Trade, and Development*, Englewood Cliffs, NJ: Prentice Hall.

Debbage, K.G. (1990) 'Oligopoly and the resort cycle in the Bahamas', *Annals of Tourism Research* 17,4: 513–27.

—— (1991) 'Spatial behavior in a Bahamian resort', *Annals of Tourism Research* 18,2: 251–68.

—— (1992) 'Tourism oligopoly is at work', *Annals of Tourism Research* 19: 355–59.

—— (1994) 'The international airline industry: globalization, regulation, and strategic alliances', *Journal of Transport Geography* 2,3: 190–203.

Deitrick, S., Beauregard, R.A. and Kerchis, C.Z. (1999) 'Riverboat gambling, tourism, and economic development', in D.R. Judd and S.S. Fainstein (eds) *Places to Play*, New Haven, CT: Yale University Press.

Del Roso, L. (1996) 'Sunmakers eyes farther horizons since buying Maupintour', *Travel Weekly* 55,19: 22.

Delaney-Smith, P. (1987) 'The tour operator: new and maturing business', in A. Hodgson (ed.) *The Travel and Tourism Industry: Strategies for the Future*, New York: Pergamon Press, 94–106.

Dev, S.D. and Klein, S. (1993) 'Strategic alliances in the hotel industry', *The Cornell HRA Quarterly* February: 42–45.

DiBella, M. and DeNicolo, L. (1985) *Cattolica*, Rimini: Maggioli Editiore.

Dicken, P. (1986) *Global Shift: Industrial Change in a Turbulent World*, London: Harper and Row.

—— (1992) *Global Shift: The Internationalization of Economic Activity*, London: Chapman Paul Publishing.

Din, K.M. (1992) 'The "involvement stage" in the evolution of a tourist destination', *Tourism Recreational Research* 17,1: 10–20.

Doeringer, P.B., Terkla, D.G. and Topakian, G.C. (1987) *Invisible Factors in Local Economic Development*, New York: Oxford University Press.

Donaghu, M.T. and Barff, R. (1990) 'Nike just did it: international subcontracting and flexibility in athletic footwear production', *Regional Studies* 24: 537–57.

Doxey, G.V. (1975) 'A causation theory of visitor–resident irritants: methodology and research inferences', in *Travel and Tourism Research Association Sixth Annual Conference Proceedings*, San Diego, California, 195–98.

Drennan, M.P. (1992) 'Gateway Cities: The Metropolitan Sources of US Producer Service Exports', *Urban Studies* 29,2: 217–35.

Drucker, P.F. (1992) 'The new productivity challenge', *Harvard Business Review* 696: 69–79.

Duncan, S.S. and Goodwin, M. (1985a) 'Local economic policies: local regeneration or political mobilisation', *Local Government Studies* 11,6: 75–96.

—— (1985b) 'The local state and local economic policy: why the fuss?' *Policy and Politics* 13: 27–253.

—— (1988) *The Local State and Uneven Development*, Cambridge: Polity Press.

Dunford, M. and Benko, G. (1991) 'Neo-Fordism or post-Fordism? Some conclusions and further remarks' in G. Benko and M. Dunford (eds) *Industrial Change and Regional Development*, London: Belhaven Press, 286–305.

Dunham, P.J. (1997) 'Reconceptualizing manufacturing-service linkages: a realist approach', *Environment and Planning A* 29: 349–64.

Dunning, J.H. and McQueen, M. (1982a) 'The eclectic theory of the multinational enterprise and the international hotel industry', in A.M. Rugman (ed.) *New Theories of the Multinational Enterprise*, New York: St Martin's Press, 79–106.

Dunning, J.H. and McQueen, M. (1982b) 'Multinational corporations in the international hotel industry', *Annals of Tourism Research* 9: 69–90.

Eagles, P.J. and Cascagnette, J.W. (1995) 'Canadian ecotourists: who are they?', *Tourism Recreation Research* 20,1: 22–28.

Eber, S. (ed.) (1992) *Beyond the Green Horizon: Principles for Sustainable Tourism*, Surrey: WWF.

Echtner, C.M. (1995) 'Entrepreneurial training in developing countries', *Annals of Tourism Research* 22,1: 119–34.

Economic Classification Policy Committee (1993) *Conceptual Issues. Issues Paper No. 1*, Washington, DC: GOP http://blue.census.gov/epcd/naics/issues1 (6 April 1997).

—— (1994a) *Economic Concepts Incorporated in the Standard Industrial Classification Industries of the United States. Report No. 1*, Washington, DC: GOP http://blue.census.gov/epcd/naics/ecpcrpt1 (6 April 1997).

—— (1994b) *Services Classifications. Issues Paper No. 6*, Washington, DC: GOP http://blue.census.gov/epcd/naics/issues1 (6 April 1997).

Economic Council of Canada (1990) *Good Jobs, Bad Jobs: Employment in the Service Economy*, Ottawa: Minister of Supply and Services Canada.

The Economist (1993a) 'The final frontier', 20 February: 63.

—— (1993b) 'Tour operators: rites of summer', 21 August: 45–46.

—— (1995) 'Faulty holiday towers', 336: 13.

—— (1996) 'Amex and Microsoft: tripping out on the web', 3 August: 54–55.

—— (1997) 'Freedom in the air', 5 April: 64–65.

Economist Intelligence Unit (EIU) (1975) 'Currency changes, exchange rates and their effects on tourism', *International Tourism Quarterly* Special Article No. 18,4: 34–45.

Edgell, L.D. (1993) *World Tourism at the Millennium*, Washington, DC: USTTA, US Department of Commerce.

Edwards, A. (1985) *International Tourism Development and Forecasts*, London: The Economist Intelligence Unit.

Eisinger, P.K. (1988) *The Rise of the Entrepreneurial State*, Madison, WI: University of Wisconsin Press.

Ellig, J. (1991) 'Computer reservation systems, creative destruction and consumer welfare: some unsettled issues', *Transportation Law Journal* 19,2: 287–307.

Employment and Immigration Canada (1994) 'Terms of reference for a human resources study of the Canadian accommodation industry', mimeo, Ottawa: EIC.

English, P. (1986) *The Great Escape: An Examination of North–South Tourism*, Ottawa: The North–South Institute.

Eurostat (1993) *Portraits of the Regions*, Luxembourg: Office for Official Publications of the European Community.

Executive Office of the President (1972) *Standard Industrial Classification Manual*, Washington, DC: GPO.

—— (1987) *Standard Industrial Classification Manual*, Washington, DC: GPO.

Fainstein, S.S. (1990) 'The changing world economy and urban restructuring', in *Leadership and Urban Regeneration: Cities in North America and Europe*, vol. 37 of *Urban Affairs Annual Reviews*, Newbury Park, CA: Sage Publications, 31–47.

Fainstein, S.S. (1991) 'Promoting economic development: urban planning in the US and Great Britain', *Journal of the American Planning Association* 57: 22–33.

Fainstein, S.S. and Stokes, R. (1995) 'Spaces to play: the impact of entertainment development on New York City', paper presented at the annual meeting of the Association of Collegiate Schools of Planning, Detroit.

Fayos-Solá, E., Marin, E.A. and Meffert, C. (1994) 'The strategic role of tourism trade fairs in the new age of tourism', *Tourism Management* 15: 9–16.

Featherstone, M. (1991) *Consumer Culture and Postmodernism*, London: Sage.

Feiffer, M. (1985) *Going Places: the Ways of the Tourist from Imperial Rome to the Present Day*, London: Macmillan

Feldman, J. (1988) 'CRS and fair airline competition', *EIU Travel and Tourism Analyst* 2: 5–22.

—— (1992a) 'Airlines lighten the load', *Air Transport World* 298: 32–36.

—— (1992b) 'Complicated kinships', *Air Transport World* 298: 49–55.

Fesenmaier, D. and Uysal, M. (1990) 'The tourism system: levels of economic and human behavior', in J.B. Zeiger and L.M. Caneday (eds) *Tourism and Leisure: Dynamics and Diversity*, Alexandria, VA: National Recreation and Park Association, 27–35.

Fisher, A.G.B. (1935) *The Clash of Progress and Security*, London: Macmillan.

Foot, D. and Stoffman, D. (1996) *Boom, Bust and Echo*, Toronto: Macfarlane, Walter and Ross.

Fourastié, J. (1947) *Esquisse D'Une Théories Générale de L'Évolution Économique Contemporaine*, Paris.

—— (1949) *Le Grand Espoir du XXe Siècle*, Paris.

Frederick, M. (1993) 'Rural tourism and economic development', *Economic Development Quarterly* 7: 215–24.

Fretter, A.D. (1993) 'Place marketing: a local authority perspective', in G. Kearns and C. Philo (eds) *Selling Places: The City as Cultural Capital, Past and Present*, Oxford: Pergamon Press, 163–74.

Fuchs, V.R. (1968) *The Service Economy*, New York: National Bureau for Economic Research.

Gamble, P.R. (1989) 'Applications of information technology-hospitality,' in S. Witt and L. Moutinho (eds) *Tourism Marketing and Management Handbook*, Hemel Hempstead: Prentice Hall International, 202–6.

—— (1991) 'Innovation and innkeeping', *International Journal of Hospitality Management* 10,1: 3–23.

Gannon, J. and Johnson, K. (1995) 'The global hotel industry: the emergence of continental hotel companies', *Progress in Tourism and Hospitality Research* 1,1: 31–42.

Gardner, D.M. (1987) 'Product life cycle: a critical look at the literature', in M. Houston (ed.) *Review of Marketing*, Chicago: American Marketing Association, 162–94.

Garrison, C.B. and Parker, A.S. (1973) 'An entropy measure of the geographic concentration of economic activity', *Economic Geography* 49: 319–24.

Gartner, W.C. (1993) 'Image formation process', in M. Uysal and D.R. Fesenmaier (eds) *Communication and Channel Systems in Tourism Marketing*, New York: The Haworth Press, 191–215.

Gee, C.Y., Makens, J.C. and Dexter, J.L.C. (1989) *The Travel Industry*, New York: Van Nostrand Reinhold.

Gershuny, J. and Miles, I. (1983) *The New Service Economy*, London: Frances Pinter.

Gertler, M.S. (1988) 'The limits to flexibility: comments on the post-Fordist vision of production and its geography', *Transactions of the Institute of British Geographers NS* 13: 419–32.

—— (1992) 'Flexibility revisited: districts, nation states, and the forces of production', *Transactions of the Institute of British Geographers NS* 17: 259–78.

Giddens, A. (1985) *The Nation-State and Violence*, Cambridge: Polity Press.

Gilbert, D.C. (1990) 'Conceptual issues in the meaning of tourism', in C.P. Cooper (ed.) *Progress in Tourism, Recreation, and Hospitality Management*, London: Belhaven Press, 24–27.

Gilbert, E. (1939) 'The growth of island and seaside hotel resorts in England', *Scottish Geographical Magazine* 55: 16–35.

Gill, K. (1997) 'Computer reservation systems: their diffusion and implications for urban tourism development', unpublished MA thesis, Department of Geography, McGill University, Montreal.

Gluck, F.W., Kaufman, S.P. and Walleck, A.S. (1980) 'Strategic management for competitive advantage', *Harvard Business Review* 58,4: 154–61.

GM Robinson Associates (1993) *Tourism, Technology and Industrial Strategy*, Toronto: GM Robinson.

Go, F.G. (1992a) 'The role of computer reservation systems in the hospitality industry', *Tourism Management* 13,1: 22–26.

—— (1992b) 'The role of CRS in the hospitality industry', *Tourism Management* 12: 22–26.

—— (1995) 'Matching global competition: cooperation among Asian airlines', *Tourism Management* 16,1: 61–65.

Go, F.G. and Pine, R. (1995) *Globalization Strategy in the Hotel Industry*, London: Routledge.

Go, F.G. and Williams, A.P. (1993) 'Competing and cooperating in the changing tourism channel system', *Journal of Travel and Tourism Marketing* 2, 2/3: 229–48.

Go, F.G., Pyo, S.S., Uysal, M., and Mihalik, B.J. (1990) 'Decision criteria for transnational hotel expansion', *Tourism Management* 11: 297–304.

Goe, W.R. and Shanahan, J.L. (1990) 'A conceptual approach for examining service sector growth in urban economies: issues and problems in analyzing the service economy', *Economic Development Quarterly* 4,2: 144–53.

—— (1991) 'Patterns of economic restructuring in industrial-based metropolitan areas', *Urban Studies* 28,4: 559–76.

Goffee, R. and Scase, R. (1983) 'Class entrepreneurship and the service sector: towards a conceptual clarification', *Services Industries Journal* 3: 146–60.

Goodall, B. (1988) 'Changing patterns and structure of European tourism', in B. Goodall and G. Ashworth (eds) *Marketing in the Tourism Industry: The Promotion of Destination Regions*, London: Croom Helm, 18–38.

—— (1992) 'Coastal resorts: development and re-development', *Built Environment* 18,1: 5–11.

Goodall, B. and Bergsma, J. (1990) 'Destinations as marketed in tour operators' brochures', in G. Ashworth and B. Goodall (eds) *Marketing Tourism Places*, London: Routledge, 170–92.

Gooding, E.G.B. (1971) 'Food production in Barbados with particular reference to tourism', in G.V. Doxey (ed.) *The Tourist Industry in Barbados*, Kitchener: Dusco Graphics.

Goodman, R. (1979) *The Last Entrepreneurs*, New York: Simon & Schuster.

Goodwin, M. (1993) 'The city as commodity: the contested spaces of urban development', in G. Kearns and C. Philo (eds) *Selling Places: The City as Cultural Capital, Past and Present*, Oxford: Pergamon Press, 145–62.

Gormsen, E. (1981) 'The spatio-temporal development of international tourism: attempt at a centre–periphery model', *La Consommation d'Espace par le Tourisme et sa Preservation*, Aix-en-Provence: CHET.

Gray, H.P. (1970) *International Travel – International Trade*, Lexington: Health Lexington Books.

Greenfield, H.I. (1966) *Manpower and the Growth of Producer Services*, New York: Columbia University Press.

Greiner, N. (1994) 'Inside running on Olympic bid', *The Australian* 19 September: 13.

Grekin, J. (1994) 'Understanding the community level impacts of tourism: the case of Pond Inlet', unpublished MA thesis, Department of Geography, McGill University.

Grekin, J. and Milne, S. (1996) 'Community-based tourism: the experience of Pond Inlet', in R.W. Butler and T.Hinch (eds) *Tourism and Native Peoples*, London: Routledge, 76–106.

Grenadier, (1995) 'The persistence of real estate cycles', *Journal of Real Estate Finance and Economics* 2: 95–119.

Grimes, P. (1991) 'High tech: high touch or high anxiety', *Cornell HRA Quarterly* October: 36–43.

Guerrier, Y. (1986) 'Hotel manager – an unsuitable job for a woman?' *The Service Industries Journal* 6,2: 227–40.

Gunn, C.A. (1994) *Tourism Planning* 2nd edn, New York: Taylor & Francis.

Hackbart, M.M. and Anderson, D.A. (1975) 'On measuring economic diversification', *Land Economics* 51: 374–78.

Hall, C.M. (1992) *Hallmark Tourism Events: Impacts, Management and Planning*, Chichester: Belhaven.

—— (1994) *Tourism and Politics: Policy, Power and Place*, Chichester: John Wiley.

—— (1995) *Introduction to Tourism in Australia: Impacts, Planning and Development*, 2nd. edn, South Melbourne: Longman Australia.

—— (1996) 'Mega-events and their legacies', in P. Murphy (ed.) *Quality Management in Urban Tourism*, New York: John Wiley & Sons, 77–89.

—— (1997) 'Geography, marketing and the selling of places', *Journal of Travel and Tourism Marketing* 6, 3/4: 61–84.

Hall, C.M. and Jenkins, J. (1995) *Tourism and Public Policy*, London: Routledge.

Hall, C.M., Jenkins, J. and Kearsley, G. (eds) (1997) *Tourism Planning and Policy in Australia and New Zealand: Cases, Issues and Practice*, Sydney: Irwin.

Hall, D.R. (1991) *Tourism and Economic Development in Eastern Europe and the Soviet Union*, London: John Wiley & Sons.

Hall, D. and Kinnaird, V. (1994) 'A note on women travelers', in V. Kinnaird and D. Hall (eds) *Tourism: A Gender Analysis*, London: John Wiley & Sons, 188–209.

Hansen, N. (1992) 'Competition, trust and reciprocity in the development of innovative regional milieux', *Papers in Regional Science* 71,2: 95–105.

Harmon, M.M. and Mayer, R.T. (1986) *Organization Theory for Public Administration*, Glenview: Scott, Foreman and Company.

Harrington, J.W. and Warf, B. (1995) *Industrial Location: Principles, Practice, and Policy*, Routledge: London.

Harrison, B. (1994) *Lean and Mean: The Changing Landscape of Corporate Power in the Age of Flexibility*, New York: Basic Books.

Harrison, B. and Bluestone, B. (1988) *The Great U-Turn*, New York: Basic Books.

Harrison, D. (1994) *Tourism and the Less Developed Countries*, London: John Wiley & Sons.

Harrison, L. and Johnson, K. (1992) *UK Hotel Groups Directory 1992/93*, London: Cassell.

Hartshorn, T. and Alexander, J. (1988) *Economic Geography*, Englewood Cliffs, NJ: Prentice Hall.

Harvey, D. (1987) 'Flexible accumulation through urbanization: reflections on "Post-Modernism" in the American city', *Antipode* 19: 260–86.

—— (1988) 'Voodoo cities', *New Statesman and Society* 30 September: 33–35.

—— (1989a) *The Condition of Postmodernity*, Oxford: Basil Blackwell.

—— (1989b) 'From managerialism to entrepreneurialism: the transformation in urban governance in late capitalism', *Geografiska Annaler* 71B: 3–17.

—— (1989c) 'The geographical and geopolitical consequences of the transition from Fordism to Flexibility', in G. Sternlieb and J. Hyde (eds) *America's New Market Geography*, New Brunswick, NJ: Rutgers, 101–34.

Haywood, K.M. (1986) 'Can the tourist area life cycle be made operational?', *Tourism Management* 7: 154–167.

—— (1990) 'A strategic approach to managing technology', *Cornell HRA Quarterly* May: 39–45.

—— (1992) 'Identifying and responding to challenges posed by urban tourism', *Tourism Recreation Research* 17: 9–23.

Healy, M.J. and Ilbery, B.W. (1990) *Location and Change: Perspectives on Economic Geography*, Oxford: Oxford University Press.

Heape, R. (1983) 'Tour operating planning in Thomson Holidays UK. The use of research', *Tourism Management* 4: 245–52.

Hennessy, S. (1994) 'Female employment in tourism development in south-west England', in V. Kinnaird and D. Hall (eds) *Tourism: A Gender Analysis*, London: John Wiley & Sons, 35–51.

Hepworth, M.E. (1987) 'The Information City', *Cities* 4: 253–62.

—— (1989) *Geography of the Information Economy*, London: Belhaven Press.

—— (1991) 'Information technology and the global restructuring of capital markets', in S.D. Brunn and T.R. Leinbach (eds) *Collapsing Space and Time: Geographical Aspects of Communications and Information*, London: Harper Collins Academic.

Herbig, P., Golden, J.E. and Dunphy, S. (1994) 'The relationship of structure to entrepreneurial and innovative success', *Marketing Intelligence and Planning* 12,9: 37–48.

Herron, J. (1993) *After Culture: Detroit and the Humiliation of History*, Detroit: Wayne State University Press.

Hicks, L. (1990) 'Excluded women: how can this happen in the hotel world?' *The Service Industries Journal* 10,2: 348–63.

304

Hills, T.L. and Lundgren, J. (1977) 'The impact of tourism in the Caribbean: a methodological study', *Annals of Tourism Research* 4,3: 248–67.

Hine, D. (1993) *Governing Italy: The Politics of Bargained Pluralism*, Oxford: Clarendon Press.

Hobson, J.S.P. (1993) 'Analysis of the US cruise line industry', *Tourism Management* 14,6: 453–62.

Hodierne, T. and Botterill, D. (1993) 'The EC travel directive – for better or worse?', *Tourism Management* 14: 331–34.

Hogwood, B. and Gunn, L. (1984) *Policy Analysis for the Real World*, Oxford: Oxford University Press.

Holcomb, B. (1993) 'Revisioning place: de- and re-constructing the image of the industrial city', in G. Kearns and C. Philo (eds) *Selling Places: The City as Cultural Capital, Past and Present*, Oxford: Pergamon Press, 133–43.

Hollinshead, K. (1993) 'The truth about Texas', unpublished Ph.D. thesis, College Station Texas: Texas A & M University.

Hoover, E.M. and Giarratani, F. (1984) *An Introduction to Regional Economics*, 3rd edn, New York: Alfred A. Knopf.

Horwath and Horwath Ltd (1987) *The Hotel Industry in the 21st Century*, London: Horwath and Horwath.

Hotels (1992) 'Special report – annual hotel industry review' 26.

—— (1996) 'Special report – annual hotel industry review' 30.

Hudman, E.L. and Davis, J.A. (1994) 'World tourism markets: changes and patterns', in *Tourism: The Economy's Silver Lining*, TTRA, 25th conference proceedings, Bal Harbour, FL, 127–45.

Hughes, G. (1991) 'Conceiving of tourism: a comment arising from Prentice R. (1990)', *Area* 23,3: 263–67.

Hughes, H.L. (1984) 'Government support for tourism in the UK: a different perspective', *Tourism Management* 5,1: 13–19.

Hull, J. (1996) 'Using the Internet to promote a sustainable tourism industry: a case study of the Lower North Shore of Quebec', paper presented at the International Geographical Union, The Hague, 4–10 August.

Husbands, W.C. (1983) 'Tourist space and touristic attraction: an analysis of the destination choices of European travelers', *Leisure Sciences* 5,3: 289–307.

Illeris, S. (1989) *Services and Regions in Europe*, Aldershot: Avebury.

Industry Science and Technology Canada (1991) *Canada's Service Economy: Facts and Figures*, Ottawa: ISTC.

Inskeep, E. (1988) 'Tourism planning: an emerging specialization', *Journal of the American Planning Association* 54: 360–72.

International Air Transport Association (1996a) *Regulatory Developments in 1996*, Geneva: IATA.

—— (1996b) *World Air Transport Statistics*, No. 40 WATS 6/96, Geneva: IATA.

International Monetary Fund (1997) *World Economic Outlook*, Washington DC: IMF.

Ioannides, D. (1992) 'Tourism development agents: The Cypriot resort cycle', *Annals of Tourism Research* 19,4: 711–31.

—— (1994) 'The state, transnationals, and the dynamics of tourism evolution in small island nations', Ph.D. dissertation, Rutgers University, New Brunswick, NJ.

Ioannides, D. (1995) 'Strengthening the ties between tourism and economic geography: a theoretical agenda', *Professional Geographer* 47,1: 49–60.

—— (1996a) 'Tourism and the economic geography nexus: a response to Anne-Marie d'Hauteserre', *Professional Geographer* 48,2: 219–21.

—— (1996b) 'Urban tourism in the United States', paper presented at the ACP–AESOP International Conference, Toronto.

Ioannides, D. and Debbage, K. (1997) 'Post-Fordism and flexibility: the travel industry polyglot', *Tourism Management* 18,4: 229–41.

IOUTO (1974) 'The role of the state in tourism', *Annals of Tourism Research* 1,3: 66–72.

Jackson, K.T. (1985) *Crabgrass Frontier*, New York: Oxford University Press.

Jafari, J. (1982) 'The tourism market basket of goods and services: the components and nature of tourism', in T.V. Singh, J. Kaur and D.P. Singh (eds) *Studies in Tourism Wildlife Parks Conservation*, New Delphi: Metropolitan Book Company, 1–12.

—— (1983) 'Anatomy of the travel industry', *The Cornell Hotel and Restaurant Administration Quarterly* 24 May: 71–77.

—— (1989) 'Sociocultural dimensions of tourism: an English language literature review', in J. Bustrzanowski (ed.) *Tourism as a Factor of Change: a Sociocultural Study*, Vienna: Economic Coordination Centre for Research and Documentation in Social Sciences, 17–60.

Jakle, J.A. (1985) *The Tourist: Travel in Twentieth-Century North America*, Lincoln: University of Nebraska Press.

Jameson, F. (1984) 'Postmodernism or the cultural logic of capitalism', *New Left Review* 146: 53–93.

Jefferson, A. and Lickorish, L. (1988) *Marketing Tourism*, Harlow: Longman.

Jeffries, D. (1989) 'Selling Britain – a case for privatisation?', *Travel and Tourism Analyst* 1: 69–81.

Jenkins, J.M. and Hall, C.M. (1997a) 'Tourism planning and policy in Australia', in C.M. Hall, J.M. Jenkins and G. Kearsley (eds) *Tourism Planning and Policy in Australia and New Zealand: Cases, Issues and Practice*, Sydney: Irwin, 37–48.

—— (1997b) 'Rural tourism and recreation: policy dimensions', in R.W. Butler, C.M. Hall and J.M. Jenkins (eds) *Tourism and Recreation in Rural Regions*, Chichester: John Wiley & Sons.

Jessop, B. (1992) 'Post-Fordism and flexible specialisation: incommensurable, contradictory, complementary, or just plain different perspectives?', in H. Ernste and V. Meier (eds) *Regional Development and Contemporary Industrial Response: Extending Flexible Specialisation*, London: Belhaven Press, 25–43.

Johnson, P. and Ashworth, G. (1990) 'Modeling tourism demand: a summary review', *Leisure Studies* 9: 145–60.

Johnson, P. and Thomas, B. (1992) 'The analysis of choice and demand in tourism', in P. Johnson and B. Thomas (eds) *Choice and Demand in Tourism*, London: Mansell, 1–12.

Johnston, J. (1972) *Econometric Methods*, Tokyo: McGraw-Hill.

Johnston, R.J. (1982) *Geography and the State*, London: Macmillan.

Johnston, R.J., Gregory, D. and Smith, D.M. (eds) (1986) *The Dictionary of Human Geography*, 2nd edn, Oxford: Basil Blackwell.

Jones, C.B. (1993) 'Applications of database marketing in the tourism industry', *Occasional Papers for the Pacific Asia Travel Association No. 1*, San Francisco: PATA.

Jud, G.D. and Josef, H. (1974) 'International demand for Latin American tourism', *Growth and Change* 5,1: 25–31.

Judd, D.R. (1988) *The Politics of American Cities: Private Power and Public Policy*, Glenview Il: Scott, Foresman and Company.

—— (1995) 'Promoting tourism in US cities', *Tourism Management* 16: 175–87.

Judd, D.R. and Swanstrom, T. (1994) *City Politics*, New York: HarperCollins.

Kaldor, N. (1966) *Causes of the Slow Rate of Growth in the United Kingdom*, Cambridge: Cambridge University Press.

Kellerman, A. (1985) 'The evolution of service economies: a geographical perspective', *Professional Geographer* 37: 133–43.

Kerlinger, F.N. (1986) *Foundations of Behavioral Research*, 3rd edn, New York: Holt, Rinehart and Winston.

Kim, Y. (1996) 'Development of a model to examine the determinants of demand for international hotel rooms in Seoul', unpublished Ph.D. dissertation, Virginia Polytechnic Institute and State University, Blacksburg, VA.

Kim, Y. and Uysal, M. (1997) 'The endogenous nature of price in tourism demand studies', *Tourism Analysis*, 2,1: 9–16.

Kinnaird, V., Kothari, U. and Hall, D. (1994) 'Tourism: gender perspectives', in V. Kinnaird and D. Hall (eds) *Tourism: A Gender Analysis*, London: John Wiley & Sons, 1–34.

Kirby, D. (1987) 'Convenience stores', in E. McFadyen (ed.) *Changing Face of British Retailing*, London: Newman Books, 94–102.

Kmenta, J. (1986) *Elements of Econometrics*, 2nd edn, New York: Macmillan.

Knowles, T. and Garland, M. (1994) 'Transport: the strategic importance of CRS in the airline industry', *EIU Travel and Tourism Analyst* 4: 4–16.

Knox, P. and Agnew, J. (1994) *The Geography of the World Economy*, London: Arnold.

Kondratiev, N.D. (1922) 'The long waves in economic life', *Review of Economic Statistics*, 17: 105–15.

Konrads, J. (1993) 'Melbourne Major Events Co. Ltd. An event impact assessment.' *Festival Management and Event Tourism* 1,2: 34–35.

Kotler, P. (1984) *Marketing Management: Analysis, Planning, and Control*, 5th edn, New York: Prentice-Hall.

Kotler, P., Haider, D.H. and Rein, I. (1993) *Marketing Places: Attracting Investment, Industry, and Tourism to Cities, States, and Nations*, New York: The Free Press.

Krippendorf, J. (1987) *The Holiday Makers: Understanding the Impact of Leisure and Travel*, Oxford: Butterworth Heinemann.

Lane, J-E. (1993) *The Public Sector: Concepts, Models and Approaches*, London: Sage.

Lanfant, M.F. (1993) 'Methodological and conceptual issues raised by the study of international tourism: a test for sociology', in D. Pearce and R. Butler (eds) *Tourism Research: Critiques and Challenges*, London: Routledge, 70–87.

Lapierre, J. and Hayes, D. (1994) 'The tourism satellite account', *National Income and Expenditure Accounts: Quarterly Estimates*, Second Quarter: xxxiii–lviii.

Latimer, M. (1985) 'Developing-island economies – tourism v. agriculture', *Tourism Management* 6: 32–42.

Laventhol and Horwath Ltd (1989) *Finding the right market niche – gone are the days of the multi-purpose hotel*, Toronto: Laventhol and Horwath.

Lavery, P. (1974) 'The demand for recreation', in P. Lavery (ed.) *Recreational Geography*, Newton Abbot: David and Charles, 22–48.

Law, C. (1994) *Urban Tourism*, London: Cassell.

Lazich, R.S. (ed.) (1996) *Market Share Reporter*, New York: Gale Research.

Lea, J. (1988) *Tourism and Development in the Third World*, London: Routledge.

Lee, F.C. (1994) 'An exploratory study of international tourism demand from the selected countries to Taiwan', unpublished Masters thesis, Virginia Polytechnic Institute and State University, Blacksburg, VA.

Leiper, N. (1979) 'The framework of tourism: toward a definition of tourism, tourist, and the tourist industry', *Annals of Tourism Research* 6,4: 390–407.

—— (1989) *Tourism and Tourism Systems*, occasional paper no. 1, Department of Management Systems, Massey University, Palmerston North.

—— (1990a) 'Partial industrialization of tourism systems', *Annals of Tourism Research* 17: 600–5.

—— (1990b) *Tourism Systems: An Interdisciplinary Perspective*, Occasional Paper No. 2, Department of Management Systems, Massey University, Palmerston North.

—— (1990c) 'Tourist attraction systems', *Annals of Tourism Research* 17,3: 367–84.

—— (1993a) 'Defining tourism and related concepts: tourist, market, industry, and tourism system', in M. Khan, M. Olsen, and T. Var (eds), *VNR's Encyclopedia of Hospitality and Tourism*, New York: Van Nostrand Reinhold, 539–58.

—— (1993b) 'Industrial entropy in tourism systems', *Annals of Tourism Research* 20,2: 221–25.

Leitch, C. (1989) 'Market segmentation continues to drive today's lodging markets', *Canadian Hotel and Restaurant* March: 20–21.

Leontidou, L. (1988) 'Greece: prospects and contradiction of tourism in the 1980s', in A.M. Williams and G. Shaw (eds) *Tourism and Economic Development: Western European Experiences*, London: Belhaven Press, 80–100.

Levere, J.L. (1996) 'Still life with tourists: can visitors revive the economy for Philadelphia?', *The New York Times* 5 May: 31, 33.

Levine, M.V. (1989) 'Urban redevelopment in a global economy: the cases of Montreal and Baltimore', in R.V. Knight and G. Gappertt (eds) *Cities in a Global Society*, vol. 35 of *Urban Affairs Annual Reviews*, Newbury Park, CA: Sage Publications, 141–52.

Lewis, C.C., and Chambers, R.E. (1989) *Marketing Leadership in Hospitality*, New York: Van Nostrand Reinhold.

Lindsay, P. (1992a) 'CRS supply and demand', *Tourism Management* 12: 11–14.

—— (1992b) 'New hospitality and tourism products: CRS supply and demand', *Tourism Management* 13,1: 11–14.

Littlejohn, D. and Beattie, R. (1992) 'The European hotel industry: corporate structures and expansion activity', *Tourism Management* 12: 27–33.

Long, C.D. (1940) *Building Cycles and the Theory of Investment*, Princeton, NJ: Princeton University Press.

Loucks, K.E. (1988) *Entrepreneurship Development in Third World Countries*, report no. 88-10-01A, St Catherines: Brock University Centre for Entrepreneurship.

Louviere, J.J. and Timmermans, H. (1990) 'Using hierarchical information integration to model consumer response to possible planning actions: a recreation destination choice illustration', *Environment and Planning* 22,A: 291–308.

Loveman, G. and Sengenberger, W. (1990) 'Introduction – economic and social reorganisation in the small and medium-sized enterprise sector', in W.

Sengenberger, G. Loveman and M. Piore (eds) *The Re-emergence of Small Enterprises: Industrial Restructuring in Industrialised Countries*, Geneva: International Institute for Labour Studies, 1–61.

Lowe, M. (1993) 'Local hero! An examination of the role of the regional entrepreneur in the regeneration of Britain's regions', in G. Kearns and C. Philo (eds) *Selling Places: The City as Cultural Capital, Past and Present*, Oxford: Pergamon Press, 211–30.

Lue, C., Crompton, J.L. and Stewart, W.P. (1996) 'Evidence of cumulative attraction in multidestination recreational trip decisions', *Journal of Travel Research* 35,1: 41–49.

Lueck, T.J. (1995) 'Lower budgets don't cut flow of tax breaks', *The New York Times* 5 July: A1, B2.

Lundberg, D.E., Krishnamoorthy, M. and Starvey, M.S. (1995) *Tourism Economics*, New York: John Wiley & Sons.

Lundgren, J. (1972) 'The development of tourist travel systems – a metropolitan economic hegemony par excellence', *Jahrbuch für Fremdenverkehr* 20: 86–120.

—— (1973) 'Tourist impact/island entrepreneurship in the Caribbean', conference paper quoted in Mathieson and Wall (1982).

—— (1975) 'Tourist penetration, the tourist product, entrepreneurial response', in *Tourism as a Factor in National and Regional Development*, no. 4, Peterborough: Department of Geography, Trent University, 60–70.

MacCannell, D. (1976) *The Tourist: A New Theory of the Leisure Class*, New York: Shocken Books.

McDermott, N. and Martinez M. (1989) 'Managing labor flexibility – hotel and catering', in S. Witt and L. Moutinho (eds) *Tourism Marketing and Management Handbook*, Hemel Hempstead: Prentice Hall International, 191–95.

McDougall, L. and Davis, C. (1991) 'Older Canadians: a market of opportunity', *Travel-log* 104, Ottawa: Statistics Canada.

McGovern, S. (1994) 'Future travel: be your own travel agent', *Montreal Gazette* 21 February: C8–9, 13.

McGuffie, J. (1990) 'CRS development and the hotel sector – parts 1 and 2', *EIU Travel and Tourism Analyst Nos. 1 and 2 - Hotels/Accommodation*, London: EIU.

—— (1994) 'CRS development and the hotel sector', *EIU Travel and Tourism Analyst* 2: 53–68.

McIntosh, R.W. and Goeldner, C.R. (1990) *Tourism: Principles, Practices, Philosophy*, New York: John Wiley & Sons.

McMullan, W. and Long, W.A. (1990) *Developing New Ventures: The Entrepreneurial Option*, San Diego: Harcourt Brace Jovanovich.

Mackun, S. (1976) 'Managing tourist hinterlands of San Remo and Rimini', *Proceedings of the Middle States and Division, Association of American Geographers*, East Stroudsburg, PA: Middle States Division, Association of American Geographers.

Macnaught, T. J. (1982) 'Mass tourism and the dilemmas of modernization in Pacific Island communities', *Annals of Tourism Research* 9,3: 359–81.

Maddala, G.S. (1988) *Introduction to Econometrics*, New York: Macmillan.

Mak, J., Moncur, J. and Yonamine, D. (1977) 'Determinants of visitors expenditures and visitor length of stay: a cross-section analysis of US visitors to Hawaii', *Journal of Travel Research* 15, Winter: 5–8.

Malecki, E.J. (1995) 'Flexibility and industrial districts', *Environment and Planning A* 27: 11–14.

Malmberg, A. (1994) 'Industrial geography', *Progress in Human Geography* 18: 532–40.

Mansfeld, Y. (1990) 'Spatial patterns of international tourist flows: toward a theoretical framework', *Progress in Human Geography* 14,3: 372–90.

Marsh, D. (1983) 'Interest group activity and structural power: Lindblom's politics and markets', in D. Marsh (ed.) *Capital and Politics in Western Europe*, London: Frank Cass.

Marshall, J. N. and Wood, P.A. (1992) 'The role of services in urban and regional development: recent debates and new directions', *Environment and Planning A* 24: 1255–70.

Marshall, J.N., Damesick, P. and Wood, P. (1987) 'Understanding the location and role of producer services in the United Kingdom', *Environment and Planning A* 19: 575–96.

Marshall, J.N., Wood, P., Daniels, P.W., McKinnon, A., Bachtler, J., Damesick, P., Thrift, N., Gillespie, A., Green, A. and Leyshon, A. (1988) *Services and Uneven Development*, Oxford: Oxford University Press.

Marston, P. (1996) 'Holiday firms to face inquiry', *Electronic Telegraph* (Internet), 534 http://www.telegraph.co.uk/et (8 November).

Martin, B. and Uysal, M. (1990) 'An examination of the relationship between carrying capacity and the tourism life-cycle: management and policy implications', *Journal of Environmental Management* 31: 327–33.

Martin, C.A. and Witt, S.F. (1987) 'Tourism demand forecasting models: choice of appropriate variable to represent tourists' cost of living', *Tourism Management* 8,3: 223–45.

—— (1989) 'Forecasting tourism demand: a comparison of the accuracy of several quantitative methods', *International Journal of Forecasting* 5: 7–19.

Martinese, G.W. and Gregoretti, M. (1996) *Rilevazione statistica delle imprese e die lavoratori dipendenti nel settore del commercio, alberghi e pubblici esercizi e terziario dei servizi vari nella Provincia di Rimini*, Rimini: CGIL di Rimini.

Massey, D. (1983) 'Industrial restructuring as class restructuring: production decentralisation and local uniqueness', *Regional Studies* 17: 73–90.

—— (1984) *Spatial Divisions of Labour: Social Structures and the Geography of Production*, London: Macmillan.

Mathieson, A. and Wall, G. (1982) *Tourism: economic, physical and social impacts*, London: Macmillan.

Maxwell, J.W. (1965) 'The functional structure of Canadian cities: a classification of cities', *Geographical Bulletin* (Canada) 7: 79–104.

Medlik, S. and Middleton, V.T.C. (1973) 'Product formation in tourism', in *Tourism and Marketing*, vol. 13, Berne: AIEST.

Meis, S., Joyal, S., Lapierre, J. and Joisce, J. (1996) 'The Canadian tourism satellite account – a new tool for measuring tourism's economic contribution', conference proceedings, Travel and Tourism Research Association 27th annual conference, 408–20.

Mercer, D. (1979) 'Victoria's land conservation council and the alpine region', *Australian Geographical Studies* 17,1: 107–30.

Michaud, J. (1991) 'A social anthropology of tourism in Ladakh, India', *Annals of Tourism Research* 18: 605–21.

Middleton, V.T.C. (1988) *Marketing Travel and Tourism*, Oxford: Heinemann.

—— (1989) 'Tourist product', in S.F. Witt and L. Moutinho (eds) *Tourism Marketing and Management Handbook*, Hemel Hempstead: Prentice-Hall, 573–76.

—— (1991) 'Whither the package tour?', *Tourism Management* 12: 185–92.

Mieczkowsi, Z.T. (1981) 'Some notes on the geography of tourism: a comment', *Canadian Geographer* 25,2: 186–91.

Mill, R.C. and Morrison, A.M. (1985) *The Tourism System: An Introductory Text*, Englewood Cliffs: Prentice-Hall.

Milne, S. (1987) 'Differential multipliers', *Annals of Tourism Research* 14,4: 499–515.

—— (1992) 'Tourism and development in South Pacific island microstates', *Annals of Tourism Research* 19,2: 191–212.

—— (1996) 'Travel distribution technologies and the marketing of Pacific microstates', in C.M. Hall and S. Page (eds) *Pacific Tourism*, London: Thomson International, 109–29.

—— (1997) 'Beyond the vicious cycle?: tourism, dependency and South Pacific microstates', in D.G. Lockhart and D. Drakakis-Smith (eds) *Island Tourism*, London: Pinter, 281–301.

Milne, S. and Grekin, J. (1992) 'Travel agents as information brokers: the case of the Baffin region, Northwest Territories', *The Operational Geographer* 10,3: 11–15.

Milne, S. and McMillan, C. (1997a) *The Restructuring of Russia's Hotel Industry*, Montreal: McGill Tourism Research Group Working Paper, McGill University.

—— (1997b) 'Russia's evolving hotel industry', McGill Tourism Research Group Working Paper, McGill University.

Milne, S. and Nowosielski, L. (1997) 'Travel distribution technologies and sustainable tourism development: the case of South Pacific microstates', *Journal of Sustainable Tourism* 5, 2: 131–50.

Milne, S., Tufts, S. and Graefe, P. (1997) *Regional Development and Flexible(?) Consumer Services*, Montreal: McGill Tourism Research Group Working Paper, McGill University.

Milne, S., Waddington, R. and Perey, A. (1994) 'Canadian rail freight in the 1990s: toward more flexible forms of organisation?' *Tijdschrift Voor Economische en Sociale Geografie* 2: 159–74.

Miossec, J.M. (1977) 'Un modele de l'espace touristique', *L' Espace Geographique* 6: 41–48.

Mommaas, H. and van der Poel, H. (1989) 'Changes in economy, politics and lifestyles: an essay on the restructuring of urban leisure', in P. Bramham, I. Henry, H. Mommaas and H. van der Poel (eds) *Leisure and Urban Processes: Critical Studies of Leisure Policy in Western European Cities*, London: Routledge, 254–76.

Momsen, J.M. (1986) *Linkages Between Tourism and Agriculture: Problems for the Smaller Caribbean Economies*, no. 45, Newcastle: Department of Geography, University of Newcastle.

Morley, C.L. (1994) 'The use of CPI for tourism prices in demand modeling', *Tourism Management* 15,5: 342–46.

Moshirian, F. (1993) 'Determinants of international trade flows in travel and passenger services', *The Economic Record* 69,206: 239–52.

Mowlana, H. and Smith, G. (1992) 'Trends in telecommunications and the tourism industry: coalitions, regionalism, and international welfare systems', in F. Go and D. Frechtling (eds) *World Travel and Tourism Review*, vol. 3, Oxford: CAB International, 163–67.

Mulligan, G.F. and Reeves, K.W. (1986) 'Employment data and the classification of urban settlements', *Professional Geographer* 38: 349–58.

Mullins, P. (1991) 'Tourism urbanization', *International Journal of Urban and Regional Research* 15,3: 326–42.

Murphy, P.E. (1985) *Tourism: A Community Approach*, New York: Methuen

—— (1992) 'Data gathering for community oriented tourism planning: case study of Vancouver Island, BC', *Leisure Studies* 11,1: 65–79.

—— (1994) 'Tourism and sustainable development', in W. F. Theobald (ed.) *Global Tourism: The Next Decade*, Oxford: Butterworth-Heinemann, 274–90.

Mutch, A. (1995) 'IT and small tourism enterprises: a case study of cottage-letting agencies' *Tourism Management* 16, 7: 533–39.

NAICS Committee (1995a) *Part XIII – Proposed new industry structure for air transportation, rail transportation, water transportation, truck transportation, transit and ground passenger transportation, pipeline transportation, scenic and sightseeing transportation, and support services for transportation*, Washington, DC: GOP. http://blue. census.gov/epcd/naics/ascii/naics_23.txt. (5 April 1997).

—— (1995b) *Part III – Proposed new industry structure for food services and drinking places and accommodations*, Washington, DC: GOP. http://blue.census.gov/epcd/ naics/ascii/naics_03.txt. (5 April 1997).

—— (1995c) *Part XII – Proposed new industry structure for management and support services*, Washington, DC: GOP. http://blue.census.gov/epcd/naics/ascii/naics_ 22.txt. (5 April 1997).

—— (1995d) *Part VII – Proposed new industry structure for performing arts, spectator sports and related industries; museums, sites and similar institutions, and recreation, amusement and gambling*, Washington, DC: GOP. http://blue.census.gov/epcd/ naics/ascii/naics_17.txt. (5 April 1997).

—— (1995e) *Part XII – Proposed new industry structure for rental and leasing*, Washington, D.C.: GOP. http://blue.census.gov/epcd/naics/ascii/naics_20.txt. (5 April 1997).

The New York Times (Internet) (1996a) 'Hfs to buy Caldwell Banker'. http://www. nytimes.com. (3 May).

—— (1996b) 'Airlines use of outsourcing will likely continue to grow'. http://www.nytimes.com. (19 June).

—— (1996c) 'American, British Airways expected to sign pact'. http://www. nytimes.com. (10 June).

Nordlinger, E. (1981) *On the Autonomy of the Democratic State*, Cambridge, MA: Harvard University Press.

Norwood, J.L. and Klein, D.P. (1989) 'Developing statistics to meet society's needs', *Monthly Labor Review* 112,10: 14–24.

Noyelle, T.J. and Stanback, T.M. (1984) *The Economic Transformation of American Cities*, New York: Roman and Allenheld.

O'Brien, K. (1990) *The UK Tourism and Leisure Market*, London: The Economic Intelligence Unit.

Office of Management and Budget (1994) *Notice of proposal to replace the Standard Industrial Classification (SIC) with a North American Industry Classification System*

(NAICS), Washington, DC: GOP. http://blue.census.gov/epcd/ascii/naicsfr2.txt. (5 April 1997).

Office of Management and Budget (1996) *Economic classification policy committee: standard industrial classification replacement – the North American Industry Classification System proposed industry classification structure*, Washington, DC: GOP. http://blue.census.gov/epcd/ascii/ naicsfr7.txt. (5 April 1997).

O'Leary, J., Uysal, M. and Bailie, G. (1993) 'Indexing high propensity travel in Canada', in *Expanding Responsibilities: A Blueprint for the Travel Industry*, TTRA, 24th Conference Proceedings, 62–72.

Oppermann, M. (1992) 'International tourist flows in Malaysia', *Annals of Tourism Research* 19,3: 482–500.

—— (1993) 'Tourism space in developing countries', *Annals of Tourism Research* 20,3: 535–60.

Organization for Economic Cooperation and Development (1997) *The Future of International Air Transport Policy*, Paris: OECD.

Pagano, M.A. and O'M. Bowman, A. (1995) *Cityscapes and Capital*, Baltimore: Johns Hopkins University Press.

Page, S. (1995) *Urban Tourism*, London: Routledge.

Pearce, D.G. (1989) *Tourist Development*, New York: Longman.

—— (1992) *Tourist Organisations*, Harlow: Longman.

—— (1995) *Tourism Today: A Geographic Analysis*, 2nd edn, Harlow: Longman.

Pearce, D.G. and Elliott, J.M.C. (1983) 'The trip index', *Journal of Travel Research* 22,1: 6–9.

Phelps, N.A. (1994) *Collaborative Inter-Firm Linkages and the Formation of Centralized Networks: Evidence from the Electronics and Related Industries*, Cardiff, Wales: Department of City and Regional Planning, University of Wales.

Philo, C. and Kearns, G. (1993) 'Culture, history, capital: a critical introduction to the selling of places', in G. Kearns and C. Philo (eds) *Selling Places: The City as Cultural Capital, Past and Present*, Oxford: Pergamon Press, 1–32.

Picot, W.G. (1986) *Canada's Industries: Growth in Jobs over Three Decades*, Ottawa: Supply and Services in Canada.

Piore, M. and Sabel, C. (1984) *The Second Industrial Divide – Possibilities for Prosperity*, New York: Basic Books.

Plog, S. (1974) 'Why destination areas rise and fall in popularity', *The Cornell HRA Quarterly* 14: 55–58.

—— (1991) *Leisure Travel: Making it a Growth Market Again*, New York: John Wiley & Sons.

Pohlmann, C. (1994) 'Service sector restructuring and its urban economic implications: the case of the Montreal tourism industry', unpublished MA thesis, Department of Geography, McGill University.

Poon, A. (1988a) 'Innovation and the future of Caribbean tourism', *Tourism Management* 9: 213–20.

—— (1988b) 'Tourism and information technologies – ideal bedfellows?', *Annals of Tourism Research* 15,4: 531–49.

—— (1989) 'Competitive strategies for a new tourism', in C. Cooper (ed.) *Progress in Recreation, Hospitality, and Recreation Management*, London: Belhaven Press, 91–102.

—— (1990) 'Flexible specialization and small size: the case of Caribbean tourism', *World Development* 18,1: 109–23.

Poon, A. (1993) *Tourism, Technology and Competitive Strategies*, Wallingford: CAB International.
—— (1994) 'The "new tourism" revolution', *Tourism Management* 15: 91–92.
Porter, M.E. (1980). *Competitive Strategy: Techniques for Analyzing Industries and Competitions*, New York: The Free Press.
—— (1985) *Competitive Advantage: Creating and Sustaining Superior Performance*, New York: The Free Press.
—— (1990) *Competitive Advantage of Nations*, New York: The Free Press.
Pretes, M. (1995) 'Postmodern tourism: the Santa Claus industry', *Annals of Tourism Research*, 22,1: 1–15
Purdie, J. (1992) 'How to make the best deal', *Financial Post*, Meetings and Conventions Supplement, 10 October: 48.
Putnam, R. (1993) *Making Democracy Work: Civic Traditions in Modern Italy*, Princeton: Princeton University Press.
—— (1996) 'The strange disappearance of civic America', *The American Prospect* 24: 34–48.
Pyo, S.S., Uysal., M. and McLellan, R.W. (1991) 'A linear expenditure model for tourism demand', *Annals of Tourism Research* 18,4: 443–54.
Qu, H. and Zhang, H.Q. (1996) 'Projecting international tourist arrivals in East Asia and the Pacific to the year 2005', *Journal of Travel Research* 35,1: 27–34.
Quaderni Del Circondario di Rimini (1994) Rimini: Circondario di Rimini.
Quayson, J. and Var, T. (1982) 'A tourism demand function for the Okanagan, B.C', *Tourism Management* 3 June: 108–15.
Qui, H. and Zhang, J. (1995) 'Determinants of tourist arrivals and expenditures in Canada', *Journal of Travel Research* 33,2: 43–49.
Quinn, J.R. and Gagnon, C.E. (1986) 'Will services follow manufacturing into decline?', *Harvard Business Review* 64,6: 95–105.
Quinn, J.R. and Paquette, P.C. (1990) 'Technology in services: creating organizational revolutions', *Sloan Management Review* 312: 67–78.
Randall, J.E. and Warf, B. (1996) 'Economic impacts of AAG conferences', *Professional Geographer* 48,3: 272–84.
Reimer, G.D. (1990) 'Packaging dreams: Canadian tour operators at work', *Annals of Tourism Research* 17: 501–12.
Renshaw, M.B. (1994) 'Consequences of integration in UK tour operating', *Tourism Management* 15,4: 243–45.
Richter, L.K. (1983) 'Tourism politics and political science: a case of not so benign neglect', *Annals of Tourism Research* 10: 313–35.
Richter, L.K. (1989) *The Politics of Tourism in Asia*, Honolulu: University of Hawaii Press.
Riley, C. (1983) 'New product development in Thomson Holidays UK', *Tourism Management* 4: 253–61.
Riley, M. and Dodrill, K. (1992) 'Hotel workers orientations to work: the question of autonomy and scope', *International Journal of Contemporary Hospitality Management* 41: 23–25.
Rimini & Co. (1991) *Rimini: Azienda di Promozione Turistica del Circondario di Rimini*, Rimini: Rimini & Co.
Rink, D.R. and Swan, J.E. (1979) 'Product life cycle research: A literature review', *Journal of Business Research* 78: 219–42.

Ritchie, J.R.B. (1991) 'Global tourism policy issues: an agenda for the 1990s', in D.E. Hawkins, J.R.B. Ritchie, F. Go, and D. Frechtling (eds) *World Travel and Tourism Review: Indicators, Trends and Forecasts* 1, Wallingford: CAB International, 149–58.

Ritchie, J.R.B. and Goeldner, C.R. (eds) (1987) *Travel, Tourism, and Hospitality Research*, New York: John Wiley & Sons.

Robins, K. (1991) 'Tradition and translation: national culture in its global context', in J. Corner and S. Harvey (eds) *Enterprise and Heritage: Crosscurrents of National Culture*, London: Routledge, 21–44.

Rodenburg, E.F. (1989) 'The effects of scale in economic development: tourism in Bali', in T.V. Singh, H.L. Theuns and F. Go (eds) *Towards Appropriate Tourism: The Case of Developing Countries*, Frankfurt am Main: Peter Lang, 205–25.

Roehl, W.S. and Fesenmaier, D.R. (1988a) *The Evolution of the Travel and Tourism Industry in Texas, 1974–1982*, paper presented at the Association of American Geographers Conference, Phoenix, Arizona, April 1988.

—— (1988b) *The Evolution of the Travel and Tourism Industry on the Texas Coast, 1974–1982*, paper presented at the National Recreation and Parks Association Leisure Research Symposium, Indianapolis, Indiana, October 1988.

Rogers, J.S. (1993) *Segmentation and Space: A Case Study of the US Lodging Industry*, paper presented at the 89th Annual Meeting of the Association of American Geographers, Atlanta, Georgia, April.

Rojek, C. (1985) *Capitalism and Leisure Theory*, Andover: Tavistock Publications.

Rosemary, J. (1987) *Indigenous Enterprises in Kenya's Tourism Industry*, Geneva: Unesco.

Rosen, C. (1997) 'Sabre CRS to go on line', *Business Travel News* 17 January <crosen@mfi.com>

Rugg, D. (1973) 'The choice of journey destinations: a theoretical and empirical analysis', *Review of Economic Statistics* 55: 64–72.

Rumelt, R. (1984) 'Toward a strategic theory of the firm', in B. Lamb (ed.) *Competitive Strategic Management*, Englewood Cliffs, NJ: Prentice Hall, 556–70.

—— (1987) 'Theory, strategy and entrepreneurship', in J. Teece (ed.) *The Competitive Challenge: Strategies for Industrial Innovation and Renewal*, Cambridge, MA: A.S. Ballanger, 137–59.

Ryan, C. (1991) *Recreation Tourism: A Social Science Perspective*, London: Routledge.

—— (1995) *Researching Tourist Satisfaction: Issues, Concepts, Problems*, London: Routledge.

Sadler, D. (1993) 'Place-marketing, competitive places and the construction of hegemony in Britain in the 1980s', in G. Kearns and C. Philo (eds) *Selling Places: The City as Cultural Capital, Past and Present*, Oxford: Pergamon Press, 175–92.

Sami, M. (1971) 'The determinants of the United States demand for tourism', *Dissertation Abstract International* 32,6: 2891–A.

Sayer, A. (1989) 'Postfordism in question', *International Journal of Urban and Regional Research* 13: 666–95.

Schattschneider, E. (1960) *Semi-Sovereign People: A Realist's View of Democracy in America*, New York: Holt, Rinehart and Wilson.

Scheslinger, L. and Heskett, J. (1991) 'The service-driven service company', *Harvard Business Review* 695: 71–82.

315

Schmidhauser, H.P. (1975) 'Travel propensity and travel frequency', in A.J. Burchard and L.J. Medlik (eds) *The Management of Tourism*, London: Heinemann.
—— (1976) 'The Swiss travel market and its role within the main tourism generating countries of Europe', *Tourist Review* 31: 15–18.

Schmoll, G.A. (1977) *Tourism Promotion*, London: Tourism International Press.

Schmulmeister, S. (1979) *Tourism and the Business Cycle: Econometric Models for the Purpose of Analysis and Forecasting of Short-term Changes in Demand for Tourism*, Vienna: Austrian Institute for Economic Research.

Schor, J. (1993) *The Overworked American*, New York: Basic.

Schumpeter, J.A. (1963) 'The analysis of economic change', in J.J. Clark and M. Cohen (eds) *Business Fluctuations, Growth and Economic Stabilization: A Reader*, New York: Random House, 55–67.

Scocozza, M. (1989) 'Deregulation – a recipe for prosperity', *IATA Review* 1: 8–10.

Scott, A. J. (1992) *Industrial Spaces: Flexible Production Organization and Regional Development in North America and Western Europe*, London: Pion.
—— (1993) *Technopolis*, Berkeley: University of California Press.

Shannon, C.E. and Weaver, W. (1949) *The Mathematical Theory of Communication*, Urbana, IL: University of Illinois Press.

Shaw, G. and Williams, A.M. (1990) 'Tourism, economic development and the role of entrepreneurial activity', *Progress in Tourism, Recreation and Hospitality Management* 2: 67–81.
—— (1994) *Critical Issues in Tourism: A Geographical Perspective*, Oxford: Blackwell.
—— (1997) 'The private sector: tourism entrepreneurship – a constraint or resource?', in G. Shaw and A.M. Williams (eds) *The Rise and Fall of British Coastal Resorts*, London: Cassell.

Shaw, G., Williams, A.M. and Greenwood, J. (1987) *Tourism and the Economy of Cornwall*, Exeter: Department of Geography, University of Exeter.

Sheldon, P.J. (1986) 'The tour operator industry: an analysis', *Annals of Tourism Research* 13: 349–65.
—— (1993a) 'Destination information systems', *Annals of Tourism Research* 20: 633–49.
—— (1993b) 'Forecasting tourism: expenditures versus arrivals', *Journal of Travel Research* 32,1: 13–20.
—— (1994) 'Tourism destination databases', *Annals of Tourism Research* 21: 179–81.

Sheldon, P.J. and Var, T. (1985) 'Tourism forecasting: a review of empirical research', *Journal of Forecasting* 4: 183–95.

Sinclair, M.T. (1991) 'The economics of tourism', in C.P. Cooper (ed.) *Progress in Tourism, Recreation, and Hospitality Management*, vol. 3, London: Belhaven Press.

Sinclair, M.T. and Stabler, M. (eds) (1992) *The Tourism Industry: An International Analysis*, Wallingford: CAB International.

Sirakaya, E., McLellan, R. and Uysal, M. (1996) 'Modeling vacation destination decisions: a behavioral approach', in D.R. Fesenmaier, J.T. O'Leary and M. Uysal (eds) *Recent Advances in Tourism Marketing Research*, New York: The Haworth Press, 57–75.

Sloane, J. (1990) 'Latest developments on aviation CRSs', *EIU Travel and Tourism Analyst* 4: 5–15.

Slyworthy, A.J. (1996) *Value Migration: How to Think Several Moves Ahead of the Competition*, Boston, MA: Harvard Business School Press.

Smeral, E. (1988) 'Tourism demand, economic theory and econometrics: an integral approach', *Journal of Travel Research* 26,4: 38–42.

Smith, G.V. (1990) 'The growth of conferences and incentives', in M. Quest (ed.) *Horwath Book of Tourism*, London: Macmillan, 66–75.

Smith, S.L.J. (1983) *Recreation Geography*, New York: Longman.

—— (1988) 'Defining tourism: a supply-side view', *Annals of Tourism Research* 15,2: 179–90.

—— (1991) 'The supply-side definition of tourism: reply to Leiper', *Annals of Tourism Research* 18: 312–18.

—— (1993) 'Return to the supply-side', *Annals of Tourism Research* 20,2: 226–29.

—— (1994) 'The tourism product', *Annals of Tourism Research* 21,3: 582–95.

—— (1995) *Tourism Analysis: A Handbook*, 2nd edn, New York: Longman.

Smith, S.L.J. and Meis, S. (1997) 'The Canadian Tourism Commission', *Annals of Tourism Research* 24: 481–83.

Smith, V.L. (1977) *Hosts and Guests*, Philadelphia: University of Pennsylvania Press.

Sorensen, A.D. (1990) 'Virtuous cycles of growth and vicious cycles of decline: regional economic decline in northern New South Wales', in D.J. Walmsley (ed.) *Change and Adjustment in Northern New South Wales*, Armidale: Department of Geography and Planning, University of New England.

Sorkin, M. (ed.) (1992) *Variations on a Theme Park*, New York: Noonday Press.

Specialty Travel Index (1995) 30 (Spring/Summer).

Stabler, M.J. (1991) 'Modelling the tourism industry: a new approach', in M.T. Sinclair and M.J. Stabler (eds) *The Tourism Industry: An International Analysis*, Wallingford: CAB International, 15–43.

Stallinbrass, C. (1980) 'Seaside resorts and the hotel accommodation industry', *Progress in Planning* 13: 103–74.

Stanback, T.M. (1979) *Understanding the Service Economy*, Baltimore: John Hopkins University Press.

Stanback, T.M. and Noyelle, T.J. (1982) *Cities in Transition: Changing Job Structures in Atlanta, Denver, Buffalo, Phoenix, Columbus (Ohio), Nashville, Charlotte*, Totowa, NJ: Allenheld and Osman.

Standard and Poor (1994) *Industry Surveys*, New York: Standard and Poor Corporation.

—— (1995) *Industry Surveys*, New York: Standard and Poor Corporation.

Statistics Canada (1988) *Tourism in Canada*, catalog 87–401, Ottawa.

—— (1992) *Travel-Log*, August, catalog no. 87–003, Ottawa.

Statistica del Turismo (1991) *Rimini: Agenzia Turistica Della Provincia di Rimini*, Stipaniuk,

—— (1995) *Rimini: Agenzia Turistica Della Provincia di Rimini*.

Stipaniuk, D.M. (1993) 'Tourism and technology: interactions and implications', *Tourism Management* 13 August: 267–78.

Storper, M. and Walker, R. (1989) *The Capitalist Imperative: Territory, Technology, and Industrial Growth*, New York: Basil Blackwell.

Strong, W.B. (1992) 'Statistical measurements in tourism', in M. Khan, M. Olsen and T. Var (eds) *Encyclopedia of Hospitality and Tourism*, New York: Van Nostrand Reinhold, 735–45.

Summary, R. (1987) 'Estimation of tourism demand by multivariable regression analysis', *Tourism Management* December: 317–22.

Swyngedouw, E. (1989) 'The heart of the place: the resurrection of locality in an age of hyperspace', *Geografiska Annaler* 71B: 31–42.

Tapscott, D. and Caston, A. (1993) *Paradigm Shift: the New Promise of Information Technology*, Toronto: McGraw-Hill.

Taylor, D.G. (1980) 'How to match plant with demand: a matrix for marketing', *Tourism Management* 1,1: 55–60.

—— (1996) 'A new direction', in D.R. Fesenmaier, J.T. O'Leary and M. Uysal (eds) *Recent Advances in Tourism Marketing Research*, New York: Haworth Press, 253–63.

Taylor, P.J. (1994) 'The state as container: territoriality in the modern world-system', *Progress in Human Geography* 18,2: 151–62.

Teaford, J.C. (1990) *The Rough Road to Renaissance*, Baltimore: The Johns Hopkins University Press.

Telfer, D.J. and Wall, G. (1995) 'Linkages between tourism and food production', *Annals of Tourism Research* 24,3: 635–53.

Theobald, W. (ed.) (1994) *Global Tourism: The Next Decade*, Oxford: Butterworth Heinemann.

Thomas, T. (1990) *Airline Reregulation: The Discussion in the United States*, background paper, Ottawa: Library of Canadian Parliament.

Thompson, G., O'Hare, G. and Evans, K. (1995) 'Tourism in The Gambia: problems and proposals', *Tourism Management* 16,8: 571–81.

Thompson, W.C. (1968) *A Preface to Urban Economics*, Baltimore: The Johns Hopkins University Press.

Toh, R., Rivers, M. and Witham, G. (1991) 'Frequent-guest programs: do they fly?', *Cornell HRA Quarterly* August: 46–53.

Touche Ross and Company (1975) *Tour Wholesaler Industry Study*, New York: Touche Ross.

Tourism Canada (1988) *Applications of Technology in the Tourism Industry*, Ottawa: ISTC.

—— (1990) *Tourism on the Threshold*, Ottawa: ISTC.

—— (1991) *Canadian Tourism Industry Performance in 1990*, Ottawa: ISTC.

—— (1994) *Product Distribution in the Tourism Industry: a Profile of Tour Operators and Travel Agencies in Canada*, Ottawa: Industry Canada.

Tourism Research Group (1996) *TEMPT Survey Report*, Exeter: Department of Geography, University of Exeter.

Townsend, A. (1991) 'Services and Local Economic Development', *Area* 23,4: 309–17.

Travel Weekly (1992) 'Travel agent survey 1992', 51,65 (30 April): special issue.

—— (1996) 'Big 3 lines gobbling market share; the watchword capacity', 55, 100 (16 December): 75.

Tremblay, S. (1990) 'Les systemes informatises de reservation, entire dans l'industrie touristique', *Teoros* 9,3: 14–18.

Truitt, L.J., Teye, V.B. and Farris, M.T. (1991) 'The role of computer reservation systems: international implications for the travel industry', *Tourism Management* 12: 21–36.

Tucker, K. and Sundberg, M. (1988) *International Trade in Services*, London: Routledge.

Tufts, S. and Milne, S. (1997) *Symbolic Capital in the Evolving Urban Economy: the*

Experience of Montreal Museums, Montreal: McGill Tourism Research Group Working paper, McGill University.

Um, S. and Crompton, J.L. (1990) 'Attitude determinants in tourism destination choice', *Annals of Tourism Research* 17,3: 432–48.

United Nations Center for Transnational Corporations (1990) *Negotiating International Hotel Chain Management Agreements: Primer for Hotel Owners in Developing Countries*, New York: United Nations.

Urry, J. (1987) 'Some social and spatial aspects of services', *Environment and Planning D: Society and Space* 5: 5–26.

—— (1988) 'Cultural change and contemporary holiday-making', *Theory, Culture and Society* 5: 35–55.

—— (1990) *The Tourist Gaze: Leisure and Travel in Contemporary Societies*, London: Sage.

—— (1995) *Consuming Places*, London: Routledge.

—— (1997) 'Cultural change and the seaside resort', in G. Shaw and A. Williams (eds) *The Rise and Fall of British Coastal Resorts*, London: Cassell, 102–16.

US Department of Commerce (1975) *Historical Statistics of the United States*, volumes 1 and 2, Washington, DC: Government Printing Office.

Uysal, M. (1985) 'Construction of a model which investigates the impact of selected variables on international tourist flows', in *The Battle for Marketing Share: Strategies in Research and Marketing* TTRA, 16th proceedings, Palm Springs, CA, 55–60.

Uysal, M. and Crompton, J.L. (1984) 'Demand for international tourist flows to Turkey', *Tourism Management* 5,4: 288–97.

—— (1985) 'An overview of approaches used to forecast tourism demand', *Journal of Travel Research* 23,4: 7–15.

Uysal, M. and Hagan, L. (1993) 'Motivation of pleasure travel and tourism', in M. Khan, M. Olsen and T. Var (eds) *Encyclopedia of Hospitality and Tourism*, New York: Van Nostrand Reinhold, 798–810.

Uysal, M. and McDonald, C.D. (1989) 'Visitor segmentation by trip index', *Journal of Travel Research* 27,3: 38–41.

Uysal, M. and O'Leary, J.T. (1986) 'A canonical analysis of international tourism demand', *Annals of Tourism Research* 13,4: 651–55.

Uysal, M., Fesenmaier, D. and O'Leary, J.T. (1994) 'Geographic and seasonal variation in the concentration of travel in the United States', *Journal of Travel Research* 32,3: 61–64.

Uysal, M., McDonald, C. and O'Leary, J.T. (1988) 'Length of stay: a macro analysis for cross-country skiing trips', *Journal of Travel Research* 32,3: 29–31.

Uysal, M., Oh, H. and O'Leary, J.T. (1995) 'Seasonal variation in propensity to travel in the US', *Journal of Tourism Systems and Quality Management* 1,1: 1–13.

Vanhove, N. (1980) 'Forecasting in tourism', *The Tourist Review* 3: 2–7.

Var, T. and Lee, C. (1993) 'Tourism forecasting: state-of-the art techniques', in M. Khan, M. Olsen and T. Var (eds) *Encyclopedia of Hospitality and Tourism*, New York: Van Nostrand Reinhold, 679–96.

Vellas, F. and Becherel, L. (1995) *International Tourism: An Economic Perspective*, New York: St Martin's Press.

Ville de Montreal (1993a) *Montreal, Open to Business: Economic Development Plan*, Montreal: CIDEM.

—— (1993b) *Montreal in Figures*, Montreal: Service des affaires institutionelles.

Visser, J.A. and Wright, B.E. (1996) 'Professional education in economic development', *Economic Development Quarterly* 10: 3–20.

Vlitos-Rowe, I. (1992) 'Destination databases and management systems', *EIU Travel and Tourism Analyst* 5: 84–108.

Wahab, S., Crampon, L.J. and Rothfield, L.M. (1976) *Tourism Marketing*, London: Tourism International Press.

Waldie, P. (1993) 'Hoteliers urged to close up rooms', *Financial Post* 10 March: 3.

Walker, R. (1985) 'Is there a service economy? The changing capitalist division of labour', *Science and Society* 39: 42–83.

Wall Street Journal (1996a) 'Greyhound firm cleared in Canada for new air service' (10 June): 4.

Wall Street Journal (1996b) 'Airlines' use of outsourcing will likely continue to grow' (19 June): 4.

Walle, A.H. (1996) 'Tourism and the Internet: opportunities for direct marketing', *Journal of Travel Research* 35,1: 72–77.

Watson, S. (1991) 'Gilding the smokestacks: the new symbolic representations of deindustrialized regions', *Environment and Planning D: Society and Space* 9: 59–70.

Weiler, B. and Richins, H. (1995) 'Extreme, extravagant and elite: a profile of ecotourists on Earthwatch expeditions', *Tourism Recreation Research* 20,1: 29–36.

Welihan, W. and Chon, K. (1991) 'Resort marketing trends of the 1990s', *Cornell HRA Quarterly* August: 56–59.

Wells, S. (1991) 'A proposal for a satellite account and information system for tourism', paper presented to the International Conference on Travel and Tourism Statistics (produced by J. Lapierre with assistance from S. Wells, K. Lal, K. Campbell, and J. Joisce), Ottawa.

Wernerfelt, B. and Karnami, A. (1987) 'Competitive strategy under uncertainty', *Strategic Management Journal* 8: 187–94.

West Country Tourist Board (1992) *Tourism Business Performance in the West Country*, Exeter: West Country Tourist Board.

Wheatcroft, S. (1990) 'Towards transnational airlines', *Tourism Management* 2,4: 353–58.

—— (1994) *Aviation and Tourism Policies: Balancing the Benefits*, London: Routledge.

Whitt, J.A. (1988) 'The role of performing arts in urban competition and growth', in S. Cummings (ed.) *Business Elites and Urban Development*, Albany, NY: SUNY Press.

Wilkinson, P. (1989) 'Strategies for tourism in island microstates', *Annals of Tourism Research* 16,1: 153–77

Williams, A.M. and Shaw, G. (1988) 'Tourism policies in a changing economic environment', in A.M. Williams and G. Shaw (eds) *Tourism and Economic Development: Western European Experiences*, London: Belhaven Press, 230–39.

—— (1996) *Tourism, Leisure, Nature Protection and Agri-tourism: Principles, Partnerships and Practice*, Brussels: EPE.

Williams, A.M., Shaw, G. and Greenwood, J. (1989) 'From tourist to tourism entrepreneur, from consumption to production: evidence from Cornwall, England', *Environment and Planning A* 21: 1639–53.

Williams, A.P. (1993) 'Information technology and tourism: a dependant factor for future survival', in F. Go and D. Frechtling (eds) *World Travel and Tourism Review*, vol. 3, Oxford: CAB International, 200–5.

Williams, A.V. and Zelinsky, W. (1970) 'On some patterns of international tourism flows', *Economic Geography* 46,4: 549–67.

Williams, C.C. (1996) 'Understanding the role of consumer services in local economic development: some evidence from the Fens', *Environment and Planning A* 28: 555–71.

Williams, P.W. and Gill, A. (1994) 'Tourism carrying capacity management issues', in W. Theobald (ed.) *Global Tourism: The Next Decade*, Oxford: Butterworth-Heinemann, 174–87.

Wilton, D. (1996) *A Comparison, Explanation, and Reconciliation of the Different Estimates of the Economic Significance of Tourism in Canada by the WTTC and Statistics Canada in the Tourism Satellite Account*, Ottawa: Canadian Tourism Commission.

Witham, G. (1985) 'Hotel companies aim for multiple markets', *The Cornell HRA Quarterly* 26: 39–51.

Witt, S.F. (1980) 'An abstract mode–abstract (destination) node model of foreign holiday demand', *Applied Economics* 12,2: 163–80.

Witt, S.F. and Martin, C.A. (1987) 'International tourism demand models: inclusion of marketing variables', *Tourism Management* 8,1: 33–40.

Witt, S.F. and Witt, C.A. (1995) 'Forecasting tourism demand: a review of empirical research', *International Journal of Forecasting* 41, February: 212–35.

Wood, P. (1991a) 'Flexible accumulation and the rise of business services', *Transactions of the Institute of British Geographers NS* 16: 160–72.

—— (1991b) 'Conceptualising the role of services in economic change', *Area* 23: 66–72.

Wood, R.C. (1992) *Working in Hotels and Catering*, London: Routledge.

Wood, S. (ed) (1989) *The Transformation of Work? Skill, Flexibility and the Labour Process*, London: Unwin Hyman.

Woodside, A.G. and Lysonski, S. (1989) 'A general model of traveler destination choice', *Journal of Travel Research* 27,4: 8–14.

World Tourism Organization (1985) *The Role of Transnational Tourism Enterprises in the Development of Tourism*, Madrid: WTO.

—— (1990) *Some Potential Impacts of CRS Development*, Madrid: WTO.

—— (1991) *Tourism to the Year 2000: Qualitative Aspects Affecting Global Growth*, Madrid: WTO.

—— (1994a) *Recommendations on Tourism Statistics*, Madrid: WTO.

—— (1994b) *Yearbook of Tourism Statistics*, Madrid: WTO.

—— (1996) *Compendium of Tourism Statistics*, 16th edn, Madrid: WTO.

World Travel and Tourism Council (1995) *Travel and Tourism: Research Edition*, London: WTTC, 99–101.

—— (1996a) *The 1996-7 WTTC Travel and Tourism Report*, London: Insight Media.

—— (1996b) *Principle for Travel and Tourism National Satellite Accounting*, London: WTTC.

—— (1997a) *Air Transport and Freer World Trade*, London: WTTC.

—— (1997b) *Travel and Tourism Jobs for the Millennium*, London: WTTC.

XIII Censimento Generale Della Populazione (1991) Rome: ISTAT.

Yeung, Henry Wai-chung (1994) 'Critical reviews of geographical perspectives on business organizations and the organization of production: towards a network approach', *Progress in Human Geography* 18,4: 460–90.

Yoon, J. and Shafer, E.L. (1996) 'Models of US travel demand patterns for the Bahamas', *Journal of Travel Research* 35,1: 50–56.

Zelinsky, W. (1994) 'Conventionland USA: the geography of a latterday phenomenon', *Annals of the Association of American Geographers* 84,1: 68–86.

Zukin, S. (1988) 'The postmodern debate over urban form', *Theory, Culture and Society* 5: 431–46.

—— (1995) *The Cultures of Cities*, Cambridge, MA: Blackwell.

Zurick, D.N. (1992) 'Adventure travel and sustainable tourism in the peripheral economy of Nepal', *Annals of the Association of American Geographers* 82,4: 608–28

INDEX